Shoulders of Giants

Copyright © 2025 Jared Flatow

All rights reserved. No part of this publication may be reproduced, distributed, or transmitted in any form or by any means, including photocopying, recording, or other electronic or mechanical methods, without the prior written permission of the publisher.

ISBN: 979-8-9930534-0-0

Library of Congress Control Number: 2025920394

Published by Quasi Convex Union, San Francisco, California

Cover by Carl Flatow

Shoulders of Giants

The story of electromagnetic theory

Jared Flatow

Quasi Convex Union

2025

For my family
who lift me on their shoulders

*Nos esse quasi nanos, gigantium humeris insidentes,
ut possimus plura eis et remotiora videre.*

*We are as dwarfs upon the shoulders of giants,
so that we may see more, and farther, than they.*

—John of Salisbury, Metalogicon (1159)
after Bernard of Chartres

Contents

Preface 1
Echoes of Ancient Currents 3
Threads of Logic 15
Intuition Leads the Way 47
The Birth of a New Language 58
A Most Natural Symmetry 102
Ripples Through Reality 138
Afterword 163
Indices 166
    Terms 166
    People 170
    Other 173
    Publications 175

Preface

The spark for this book was ignited by my father's inquisitive nature and my lifelong fascination with the elegance of mathematics. It is my hope that, through these pages, I can share the beauty and wonder of mathematics beginning with calculus, a subject too often obscured by dense notation and abstract concepts.

My father has always been the kind of person who loves understanding how things work — he's fascinated by the forces that power our world, the sparks that light our homes, the magnets that steer our compasses. But when it comes to the language of mathematics, especially calculus, it's never really clicked for him. He knows it's valuable — he even encouraged me to study it — but it's always felt like someone else's tool, a key to understanding that he never quite had in his hand.

And he's not alone. For so many people, math feels abstract and disconnected from the physical world they interact with every day. The idea of using math to model reality, to predict outcomes, or even to see hidden patterns in nature — it's not that they think it's impossible; it's just never been part of their experience. This book is my attempt to bridge that gap, to show how math can bring the physical world into sharper focus. Not by slogging through dry formulas or dense equations, but by unraveling the mystery of one of the most mystifying forces of nature: electromagnetism.

For millennia, civilizations around the world have observed the phenomena of electricity and magnetism: the jolt of static on a cold day, the pull of a lodestone. Some may have possibly understood these phenomena in profound ways that have been lost to history. Legends and archaeological finds hint at knowledge that might parallel or even surpass modern understanding.

In the 19th century, visionaries like Michael Faraday and James Clerk Maxwell brought electromagnetism into the modern consciousness, eventually articulating a unified theory of it using the language of mathematics. Their work revealed an unseen, interconnected structure of reality — one so profound that it forms the backbone of our modern technological world.

Faraday was a brilliant experimenter who intuitively uncovered a working model of electromagnetism. Maxwell's genius was in communicating it. Using the tools of calculus, he gave us a precise, universal way to understand and share the rules of electromagnetism — a way to describe how the invisible forces around us behave and interact.

This book is about describing the world through the language of mathematics. It's about learning to use math not as an abstract puzzle, but as a lens to see deeper into nature. Along the way, we'll study the fundamentals of calculus — not in the way you might have encountered them in school, but in a way that's grounded in reality and built to reveal something astonishing. By the end, you'll not only understand Maxwell's equations; you'll understand why they matter, how they describe the hidden order of the universe, and why math is a tool you can wield to explore it.

This book is for my dad, and for my son, and for anyone who shares our curiosity. It's for those who want to not only understand the mechanisms of the world but also appreciate the profound elegance hidden within the mathematical descriptions of those mechanisms.

Jared Flatow
January 2025

Echoes of Ancient Currents

The enigmatic forces of electricity and magnetism have captivated human imagination for millennia. From the crackle of lightning in a stormy sky to the magical pull of a lodestone on iron, these phenomena are shrouded in mystery, often attributed to divine or supernatural forces. As far as history can recall, electricity and magnetism remained separate, seemingly unrelated forces, their intricate dance hidden from human comprehension.

The Greeks used the word *electron* (ἤλεκτρον), to refer to amber. This word forms the root of the modern term *electricity*. Thales of Miletus, around 600 BCE, noted that amber, when rubbed with fur, could attract lightweight particles such as straw, feathers, and other small bits of matter. The word *magnet* comes from the ancient Greek *magnítis líthos* (μαγνήτης λίθος), meaning *Magnesian stone*, a type of iron ore found in the region of Magnesia in Greece, which naturally displayed the ability to attract iron. The term *lodestone* (lode meaning 'way' or 'journey' in Middle English) emerged late in the medieval period, specifically referring to the ore's use in navigation.

Thales remarked that both amber and the Magnesian stone possess a kind of life because they move other objects. Plato, in his dialogue *Ion* (circa 380 BCE), referred to Magnesian stone, noting "this stone not only attracts iron rings, but it imparts to them a similar power of attracting other rings; so that you may see sometimes a long chain of iron rings and other iron substances hanging one from the other; and this is one of the properties in which the stone is peculiar". Aristotle repeated some of these observations in his own writings (circa 350 BCE). In his encyclopedia, *Naturalis Historia* (circa 77 CE), Pliny the Elder, a Roman author, naturalist, and natural philosopher, calls magnetite "a stone endowed with both sense and hands". Pliny marvels at tales of 'male' and 'female' magnets, and even notes a mineral that repels iron. He lists powdered magnetite among approved remedies for eye inflammations and burns.

The magnetic compass was introduced to Europe through the Islamic world, where it was likely brought from China. The earliest European mention of the compass is in *De naturis rerum* by the English writer Alexander Neckam in 1187 CE, where he describes its use by sailors to navigate during cloudy conditions. The technology was refined over the centuries in Europe, with the dry compass developed in the 13th century and the liquid-filled compass in the 14th century, significantly improving the accuracy and reliability of maritime navigation.

Long before the compass reached Europe, China had been exploring magnetism for centuries. The earliest surviving Chinese mention appears in the 4th-century BCE *Master Lü's Spring and Autumn Annals* (呂氏春秋, Lü Shi Chun Qiu), where a magnetic stone (磁石, cí shí) is described in the context of geomancy. In this tradition, later called *feng shui* (风水), orientation was thought to shape health, fortune, and harmony between Heaven, Earth, and humanity. Direction was not just a matter of location — it was a matter of cosmic balance. By the Han Dynasty (206 BCE – 220 CE), magnetic stones were described as naturally aligning with the Earth's fundamental forces. Later accounts spoke of 'south-pointing' devices that could pivot and settle in a steady direction, though no confirmed examples survive from this period. Whether these instruments physically existed or not, the idea reflected a consistent belief: magnetism was a visible trace of an invisible natural order.

This symbolic role began to change during the Song Dynasty (960–1279 CE), when the same directional property found a new life at sea. In 1088 CE, the polymath Shen Kuo described in his *Dream Pool Essays* (梦溪笔谈, Meng Xi Bi Tan) how a magnetized needle could be used aboard ships to find direction. He also noted that the needle did not point exactly to true north but was offset slightly — a phenomenon now known as *magnetic declination*, caused by the difference between the Earth's magnetic north and its geographic north pole. Three decades later, in 1119 CE, Zhu Yu — a former ship inspector and author of *Pingzhou Table Talks* (萍洲可談, Pingzhou Ketan) — gave a clear, practical account: sailors in the South China Sea routinely used a floating magnetized needle to set their course when clouds, fog, or heavy rain hid the sun and stars for days at a time. Unlike celestial navigation, which depended on clear skies and a skilled eye, the compass worked day or night, in any weather, making it the most dependable navigational aid available.

The Olmecs, among the earliest known civilizations of Mesoamerica, were working magnetite at least three thousand years ago. Archaeological excavations have revealed finely polished magnetite mirrors dating to around 1000 BCE, their dark, glassy surfaces ground with remarkable precision. What the Olmecs saw in these reflections is unknown. With no surviving written records, even the name they gave the material — if there was one — has been lost. Some scholars suggest the mirrors were used in ritual or divination, perhaps to glimpse visions or symbolic images, though there is little direct evidence linking their use to the stone's magnetic properties. Modern understanding comes almost entirely from the artifacts themselves, silent witnesses to a culture whose voice is gone.

Another enigmatic artifact emerged in 1938 on the outskirts of Baghdad. During excavations at Khujut Rabu, workers uncovered small terracotta jars, each fitted with a copper cylinder and an iron rod. Dated to the Parthian or Sassanian periods (approximately 250 BCE to 224 CE), these objects became famous as the Baghdad Battery after Wilhelm König proposed in the mid-20th century that they were ancient galvanic cells, perhaps used for electroplating gold onto silver. The theory remains controversial, with other explanations ranging from storage containers to ritual vessels, but the design continues to invite speculation.

Egypt offers some of the most ambitious theories of all. The Great Pyramid of Giza, built more than four thousand years ago with extraordinary geometric precision, has been reimagined by a few modern authors — notably Christopher Dunn in *The Giza Power Plant* (1998) — as an ancient energy generator. This interpretation points to the pyramid's dimensions, internal layout, and the conductive qualities of its granite and limestone. While some maintain that it is a funerary monument, the pyramid's unmatched accuracy and scale ensure that such technological theories persist.

Elsewhere in Egypt, the Temple of Hathor at Dendera presents another puzzle. Bas-reliefs on its walls depict elongated, tube-like objects ending in lotus-shaped forms, from which emerges a serpent-like figure. Some theorists claim the shapes resemble oversized light bulbs, the serpent acting as a filament, with the lotus base connected to a supposed power source. The reliefs remain among the most unusual motifs in Egyptian temple art and continue to draw attention for their suggestive design.

In Ancient Egyptian myth (circa 3150-30 BCE), lightning and thunder were linked to the god Set, the embodiment of chaos, violence, and disorder. Set's eternal adversary was Horus, the falcon-headed god who represented kingship, order, and the rightful harmony of the world. Their rivalry stemmed from the mythic struggle for the throne of Egypt after the murder of Osiris, Horus's father, by Set. This cosmic feud was mirrored in natural phenomena: storms, with their blinding flashes and shattering noise, were seen as Set's assaults, met and resisted by the stabilizing power of Horus. Such storms were not random events but visible signs of an ongoing battle between disorder and balance. This belief in lightning as a divine weapon was shared in Ancient Mesopotamia (circa 3500-539 BCE), where Sumerian and Babylonian myths describe the storm-god Adad wielding bolts from the sky. Across these cultures, the flash of lightning and the roar of thunder were not seen as mere weather, but as manifestations of divine will.

In the Ancient Indian *Rigveda* (circa 1500–1200 BCE), Indra, king of the gods, wields the *vajra* (वज्र) — a weapon described as a thunderbolt. The Sanskrit word for lightning, *vidyut* (विद्युत), carries both physical and divine meaning, signifying the flash in the sky and the heavenly force behind it. In these hymns, Indra's thunderbolt is not merely destructive; it is the decisive power that strikes down demons, releases the rains, and restores cosmic order.

The Chinese *Book of Changes* (易经, I Ching, circa 1000–750 BCE) casts thunder as a dynamic agent of transformation. The text is built around 64 hexagrams — six-line figures made of broken and unbroken lines — each representing a situation or process of change. Hexagram 51, *zhèn* (震), meaning "thunder", depicts a moment of sudden shock. It describes thunder as both frightening and invigorating, the kind of disruption that shatters complacency and forces renewal. The I Ching treats such upheaval not as random destruction but as part of the cyclical balance between stillness and movement, echoing the belief that natural forces carry moral significance. Another Chinese classic, the *Book of Mountains and Seas* (山海经, Shan Hai Jing, circa 400–300 BCE), catalogs mythical beings and deities tied to the elements, including spirits of thunder and lightning whose appearances signal both danger and seasonal change.

In the Greek world, lightning was the supreme weapon of Zeus, king of the gods, who hurled it against those who defied him. The earliest surviving Greek literature, the epics of Homer (circa 850 BCE), often invokes Zeus's lightning to mark decisive divine intervention. Hesiod (circa 700 BCE), in works such as the *Theogony* and *Works and Days*, describes the forging of Zeus's thunderbolts by the Cyclopes and their role as instruments of justice and punishment. In both poets' accounts, the flash in the sky is not a natural accident but an unmistakable signal of the ruler of Olympus asserting his will.

Further north, Norse tradition gave thunder and lightning to Thor, who wielded the hammer Mjölnir. In the sagas and the *Eddas* (circa 1220 CE) — the medieval Icelandic texts that preserve much of Norse myth — each swing of the hammer could shatter giants and send flashes across the sky, a display of divine strength that defended both gods and humankind.

In the Hebrew Bible, lightning is consistently portrayed as a direct manifestation of divine presence and authority. In the *Book of Exodus* (circa 1446 BCE), thunder (קלת, kolot) and lightning (בְּרָקִים, barakim) blaze over Mount Sinai as God descends in fire to deliver the *Ten Commandments* — a scene from the covenantal narrative underscoring the holiness of the moment. The *Book of Job* (circa 600 BCE) speaks of God's control over the rain and "a way for the lightning of the thunder" — part of poetic dialogues that highlight the mysteries of God's creations, showing the vastness and the unfathomable nature of divine acts.

One of the most enduring biblical artifacts, the Ark of the Covenant, has drawn modern speculation about its possible electrical nature. Described as a gold-covered chest carried on poles, the Ark's conductive materials and strict handling instructions have been compared to procedures for managing high-voltage equipment. Some theorists point to the biblical account of Uzzah, who died upon touching the Ark, as resembling an electrical shock.

No matter the interpretation, across these cultures appears a global fascination with and reverence for electrical phenomena, often intertwined with divine or supernatural powers. From ancient hymns and mythologies to classical philosophies and early scientific inquiries, the awe inspired by thunder and lightning transcends cultures, encapsulating a universal human experience of these powerful phenomena.

As human history progressed, so did the questions surrounding electricity and magnetism. Could these seemingly disparate forces be connected? Could there be a unifying principle that governed their behavior? These questions lingered in the minds of scientists and philosophers for centuries, fueling a quest for understanding. The stage was set for a grand scientific drama, a mystery that would unravel over centuries of investigation, experimentation, and theoretical breakthroughs. The connection between electricity and magnetism was not apparent, and the tools to probe this connection were yet to be invented. However, the seeds of curiosity had been sown, and the pursuit of knowledge would eventually illuminate the profound relationship between these two fundamental forces of nature.

In 1269, the French mathematician and engineer Petrus Peregrinus de Maricourt produced the earliest surviving treatise devoted entirely to magnetism. His *Letters on the Magnet* (Epistola de Magnete) described a series of experiments on magnetic stones, including an unambiguous account of magnetic polarity — the observation that a magnet has two fixed points with opposing tendencies to attract or repel. Peregrinus also noted how these poles behaved when brought near one another, establishing a foundational vocabulary for the study of magnetic behavior in Europe.

In the Islamic world, meanwhile, a centuries-long intellectual movement was preserving, translating, and expanding the knowledge of earlier civilizations. Under the Abbasid Caliphate, especially in Baghdad's House of Wisdom, scholars rendered into Arabic the works of Greek philosophers, Indian mathematicians, and other sources from across the known world. These translations were not mere copies; they were often accompanied by commentaries and refinements that deepened the understanding of natural philosophy. Although explicit references to magnetic navigation in Islamic sources from this period are scarce, the scientific culture they fostered broadened the conceptual framework within which such phenomena could be studied.

The movement of knowledge from the Islamic world into Europe — through the Crusades, the translation schools of Spain during the Reconquista, and the constant interchange of goods and ideas along Mediterranean trade routes — helped shape the context in which European scholars approached magnetism. Theoretical advances in China and the Islamic world had already provided key conceptual and practical insights long before similar developments emerged in Europe. Together they formed a vast, interconnected record of inquiry: a global tapestry of intellectual achievement that, piece by piece, was drawing humanity closer to understanding the hidden links between the forces of nature.

Although navigators knew that a compass needle settled in a consistent direction they called 'north', the reason for this behavior remained obscure for centuries. Petrus Peregrinus had recognized that the needle aligned toward geographic north, but he offered no explanation for why. In the centuries that followed, practical seamanship advanced faster than scientific theory.

From the 13th to the 16th centuries, improvements in compass design spread across Europe and Asia, driven by the demands of navigation. Mariners developed the dry compass and, later, the liquid-filled compass to steady the needle. The gimballed compass allowed the instrument to remain horizontal even as a ship pitched and rolled, while the compass card, marked with the points of direction, let navigators take quick readings without tracking the needle itself. These refinements made the compass more reliable at sea, but the force that caused it to point north was still explained in vague terms — often attributed to celestial alignments, hidden currents in the air, or other unseen influences.

In 1600, William Gilbert (1544-1603) published *On the Magnet* (De Magnete), a work that redefined the study of magnetism. Through extensive experiments with natural magnets and small spherical models he called *terrellas* ("little Earths"), Gilbert concluded that the Earth itself was a giant magnet, and that a compass needle aligned along its magnetic field. This was a radical departure from earlier theories that treated magnets as isolated curiosities, unrelated to the planet as a whole. By linking the compass's behavior to terrestrial magnetism, Gilbert provided a coherent scientific explanation for why the instrument worked.

The three centuries between Peregrinus and Gilbert had been marked by steady, incremental refinements in navigational tools rather than breakthroughs in magnetic theory. Gilbert's work shifted the balance: navigation was no longer solely an empirical craft, but one grounded in a physical principle that could be tested and demonstrated.

Gilbert is also remembered for clearly distinguishing magnetism from static electricity. He introduced the term *electricus* to describe materials which, when rubbed, could attract light objects such as feathers — a phenomenon known today as *electrostatics*. He noted that such electric effects produced only attraction, whereas magnets could both attract and repel. To explain the temporary nature of electric attraction, Gilbert proposed the existence of a subtle emanation he called *effluvium*.

Modern science recognizes that both electrical and magnetic forces can produce attraction and repulsion, and that Gilbert's account was incomplete. Yet in his time, the two phenomena were often blurred together, with many assuming they shared a common, supernatural cause. By separating them conceptually, Gilbert gave future investigators a clearer target for experimentation, laying the groundwork for the eventual development of electricity and magnetism as distinct but related fields.

Building on William Gilbert's foundational work in the early 17th century, the study of electricity and magnetism continued to advance through a mix of inventive apparatus, careful observation, and persistent experimentation.

In the latter half of the 17th century, Otto von Guericke (1602-1686) invented an electrostatic generator — a globe of solid sulfur mounted on an axle. When the sphere was rotated and rubbed by hand, it produced visible sparks and a noticeable electrical charge. This apparatus gave a practical and repeatable means of generating static electricity, extending the study of electrical phenomena well beyond Gilbert's original observations. Around the same time, Robert Boyle (1627-1791) demonstrated that electric attraction could occur even in a vacuum, showing that air was not necessary for the transmission of electric forces. This challenged older assumptions that some subtle property of air was required for electrical effects to occur.

In the 1720s, Stephen Gray (1666-1736) proved that static electricity could be transmitted over significant distances while retaining enough strength to do work. His experiments often used materials readily available at the time. In one of his best-known demonstrations, Gray constructed a 50 meter line of wet hemp twine, suspended at intervals by silk threads to prevent it from grounding. At one end, he generated a charge using a glass tube vigorously rubbed to create static electricity, then brought the charged tube near or into contact with the twine. At the opposite end, small pieces of gold leaf or other lightweight materials were positioned. When the twine was charged, these materials moved toward it or clung to it, proving that the electrical force had traveled the entire length. This experiment showed that electricity could be conveyed over long distances without substantial loss of effect — an insight that challenged prevailing ideas about its range and behavior.

In the 1730s, Charles François du Fay (1698-1739) carried the work further, discovering that not all electrical charges were the same. Using common materials such as glass, resin, silk, and fur, he observed that two objects charged by the same process — for example, two pieces of glass rubbed with silk — would repel one another. But if one was charged by rubbing glass with silk and another by rubbing resin with fur, they would attract. From this, du Fay concluded that there were two distinct kinds of electrical fluid, which he named *vitreous electricity* (from glass) and *resinous electricity* (from resinous substances). This distinction was crucial: it revealed that electrostatic forces could cause both attraction and repulsion.

In 1745, two scientists working independently — Ewald Georg von Kleist (1700-1748) in Pomerania and Pieter van Musschenbroek (1692-1761) at the University of Leiden — discovered a device capable of storing and releasing a substantial electrical charge. This device became known as the *Leyden jar*. Von Kleist's version used a medicine bottle filled with alcohol, with a nail suspended inside. He found that when a charge from a static electricity source was applied, the bottle could hold that charge until it was discharged by contact with a conductor. Soon after, van Musschenbroek made a similar discovery using water in a glass jar connected by a wire to a static generator. When touched by a conductor, the jar released a powerful spark.

The Leyden jar was a form of *capacitor* — a vessel for storing and releasing electrical energy on demand. Early models relied on a glass jar as the insulating medium, with the interior either filled with liquid or lined with metal foil to act as the inner conductor. A metal rod or wire passed through the lid, dipping into the conductive material inside, and was exposed at the top to receive charge from a friction machine. Later refinements added metal foil to the outer surface of the jar, connected to a grounding path. The jar retained its charge until the inner and outer conductors were connected, at which point the stored energy was released in an instant. This ability to store large amounts of charge transformed electrical research, enabling experiments that required sudden, intense discharges.

Benjamin Franklin (1706-1790) conducted numerous experiments with electricity in the 1740s and 1750s in which he used the Leyden jar extensively. His most famous — the kite experiment of June 1752 — tested whether lightning and laboratory-generated static electricity were the same. Flying a kite during a thunderstorm, Franklin attached a metal key to the wet kite string, with the key connected to a Leyden jar. The wet string conducted the atmospheric charge down to the key, where Franklin observed sparks jumping to his knuckle. Because such sparks were identical in appearance and behavior to those produced by a Leyden jar charged by a friction machine, Franklin concluded that the electrical discharge of lightning and laboratory static electricity shared the same nature.

Franklin advanced a 'single-fluid' theory of electricity, replacing Charles François du Fay's earlier 'two-fluid' model of vitreous and resinous charges. In Franklin's view, electricity was a single fluid that could exist in excess (positive charge) or deficit (negative charge) within a material. This framework made it easier to explain phenomena such as electrical discharges and grounding. He reasoned that the Earth was a vast reservoir of neutral charge. A grounded object with excess charge would discharge into the Earth, while one with a deficit would draw in charge from it. Lightning rods, in his design, provided a direct, harmless path for atmospheric charges to reach the ground, protecting buildings from lightning strikes.

In 1759, Franz Ulrich Theodor Aepinus (1724–1802), a German-born physicist working in Russia, published *An Attempt at a Theory of Electricity and Magnetism* (Tentamen Theoriae Electricitatis et Magnetismi). Aepinus had learned of Franklin's work through translations and summaries circulating in Europe, particularly via the influential networks of scientific correspondence and the published proceedings of learned societies. His treatise constructed a unified mathematical treatment of both electricity and magnetism, using a numeric quantity to describe the electrical state of a system, and applying similar reasoning to magnetic poles. Although he did not claim that electricity and magnetism were the same physical phenomenon, he recognized that they could be described with parallel mathematical forms — a striking conceptual move at a time when no experimental evidence linked them. In effect, Aepinus's unification was formal rather than physical: he was grouping them together because of their shared properties of attraction and repulsion, and because both seemed to obey similar distance laws.

Henry Cavendish (1731–1810), working largely in private in the 1770s and 1780s, conducted exceptionally precise quantitative experiments on electricity. Using Leyden jars, pith balls, and sensitive balances, he measured the forces between charged bodies and confirmed that they varied with distance in a predictable, mathematical way. Cavendish also investigated the conductivity of materials and studied the capacity of Leyden jars with remarkable accuracy. Much of Cavendish's work, however, remained unpublished during his lifetime.

Luigi Galvani (1737–1798), an Italian physician and anatomist, began experimenting with the effects of electricity on animal tissues around 1780. While dissecting frogs, he noticed an unusual phenomenon: during a thunderstorm, when he touched the exposed nerve of a frog's leg with a pair of metal scissors, the muscle twitched. Galvani pursued the observation systematically, using metal scalpels, hooks, and wires. He found that the effect was especially pronounced when two different metals came into contact with each other and the frog's tissue. From these results, Galvani concluded that the contractions were caused by what he called *animal electricity* — a unique form of electricity generated and stored within living muscle, released upon contact with certain conductors. He believed that the muscles themselves acted as reservoirs of this vital electrical force.

Alessandro Volta (1745–1827), professor of physics at the University of Pavia and a contemporary of Galvani, was intrigued but unconvinced by the theory of animal electricity. Repeating Galvani's experiments, Volta confirmed the muscle contractions but arrived at a different interpretation. He argued that the source of the electricity lay not in the animal tissue but in the contact between the two different metals used in the experiments. This, he maintained, was *contact electricity* — electricity generated at the junction of dissimilar metals in the presence of a conductive path.

To demonstrate his theory, Volta experimented with numerous metal combinations and eventually devised a completely animal-free apparatus. In 1800, he introduced the *voltaic pile* — a type of chemical battery — consisting of alternating discs of zinc and copper, each pair separated by pieces of cardboard soaked in a saline or acidic solution. This arrangement produced a steady, continuous flow of electric current, a property unmatched by the brief, irregular discharges of friction machines or Leyden jars.

Volta's invention marked a turning point in electrical science. The voltaic pile provided a reliable and controllable source of current, enabling experiments that were impossible with more sporadic forms of electricity. It opened the door to systematic study of chemical reactions driven by electricity, the discovery of new elements, and the eventual development of electrical devices and technologies on a scale previously unimaginable.

During the 18th century, advances in the understanding of electricity often outpaced those in magnetism, yet the study of magnetic phenomena also progressed in meaningful ways. Researchers began to investigate systematically both the behavior of magnetic materials and the structure of the Earth's magnetic field.

In 1701, Edmond Halley (1656–1742), as a captain of the Royal Navy of England, produced a magnetic chart of the Atlantic Ocean, mapping the variation — or *declination* — of the compass needle at sea. Magnetic declination is the angular difference between true geographic north and the direction indicated by a compass. Halley's chart revealed that this variation changed with location, an insight of great importance for navigation. He also investigated *magnetic inclination*, or *dip* — the angle the needle makes with the horizontal plane — and encouraged systematic measurement of both properties. These observations laid the foundation for later concepts such as *isogonic lines*, which connect points of equal magnetic declination, and *isoclinic lines*, which connect points of equal inclination. As more data was gathered, it became clear that the Earth's magnetic field varied not only from place to place but also over time, revealing magnetic anomalies that would challenge later theorists to explain them.

In 1750, the English scientist John Michell (1724–1793) described the phenomenon of magnetic induction, showing that a magnetized steel bar could induce magnetism in an unmagnetized iron bar brought near it. This was a step toward understanding that magnetism could be created artificially rather than being found only in natural lodestones. The production and study of artificial magnets became increasingly common, with researchers experimenting by stroking iron bars with lodestones or arranging magnets in specific configurations to strengthen their effect.

Although the terms themselves would come later, investigators of this period were already exploring what would be called *permeability* — how readily a material can be magnetized — and *retentivity* — how well it retains magnetization after the magnetizing force is removed. They found that iron and steel displayed markedly different behaviors, with steel better at retaining magnetism but iron more easily magnetized. Nickel, cobalt, and their ores were also tested and cataloged for magnetic strength. Researchers examined the effects of temperature, discovering that heating a magnet could weaken or destroy its magnetism, and they experimented with alloying metals to alter their magnetic properties.

By the late 18th and early 19th centuries, the possibility of a connection between electricity and magnetism was increasingly discussed among natural philosophers. Observations of electrical and magnetic phenomena side by side, and occasional hints that they might influence one another, had accumulated over decades, but no experiment had yet demonstrated a direct physical link.

The connection was finally solidified in 1820, when Hans Christian Ørsted (1777–1851), professor of natural philosophy at the University of Copenhagen, discovered that an electric current could deflect the needle of a magnetic compass. The finding occurred essentially by accident, while he was preparing for a lecture demonstration. Setting up a voltaic pile with a wire laid parallel to a nearby compass, Ørsted noticed that the compass needle shifted from its north–south alignment whenever the current flowed through the wire. When the current ceased, the needle returned to its original position. Ørsted published his results later that year under the title *Experiments on the effect of a current of electricity on the magnetic needle* (Experimenta circa effectum conflictus electrici in acum magneticam). The paper circulated rapidly throughout Europe, igniting further research and experimentation in the field of electromagnetism.

In France, André-Marie Ampère (1775–1836) began investigating Ørsted's reported effect almost immediately after learning of it. He examined the forces between two parallel wires carrying electric current and observed that the wires attracted one another when the currents flowed in the same direction, and repelled when they flowed in opposite directions. Ampère extended his work to coils of wire — later called *solenoids* — and showed that when current passed through them, they exhibited the behavior of magnets. In 1826, he developed a mathematical formulation — later known as *Ampère's Law* — relating the magnetic field around an electric current to the magnitude of the current. This work provided a structured way to describe and predict the interactions between currents, offering a framework for analyzing more complex arrangements of conductors and magnetic effects.

By the early 1820s, speculation on the relationship between electricity and magnetism had become active investigation, ripe for another breakthrough in understanding. Entering this dynamic scene was Michael Faraday (1791–1867), a self-taught scientist whose unconventional background in bookbinding further sharpened his innovative edge. Building on established theories, Faraday was poised to delve deeper into the mysterious interplay between electric currents and magnetic fields. His forthcoming experiments promised not just to expand the boundaries of knowledge but also to transform human control over nature's unseen forces.

Despite what would become Faraday's groundbreaking work in electromagnetism, he lacked formal mathematical training, relying instead on his remarkable intuition and experimental skill. To fully appreciate the framework that would eventually formalize Faraday's discoveries, the development of mathematics leading up to his time should first be traced.

Threads of Logic

Logic and mathematical modeling have been central to human inquiry since antiquity, valued for their ability to bring clarity to complex problems, predict outcomes, and guide decision-making. These tools have shaped the course of civilizations, enabling societies to describe, interpret, and ultimately harness aspects of the natural world to their advantage.

The origins of logic are closely tied to the beginnings of human communication and abstract thought. While logical reasoning almost certainly existed in human minds long before it was recorded, the deliberate formalization of logic — treating it as a subject to be studied, refined, and taught — marked a turning point in how ideas could be examined and tested.

Ancient peoples often viewed their capacity for reason as a divine gift or a natural reflection of the cosmic order. In many cultures, this ability was not seen as a mere accident of evolution but as an essential part of being human — a spark imbued by the gods or woven into the fabric of creation.

Mathematics, built upon a foundation of logical principles, became a symbolic language for expressing patterns and quantitative relationships. Where logic defines the rules of valid reasoning, mathematics uses those rules to construct a rich and adaptable framework capable of modeling phenomena ranging from the motions of planets to the flow of water.

To model is to create a representation — whether a set of rules, a system of equations, or a physical replica — that captures the essential features of something, allowing it to be studied and understood. The English word *model* comes from the Italian *modello*, meaning a small-scale representation, ultimately derived from the Latin *modus*, meaning 'measure', 'manner', or 'way'. The sense of a "small-scale representation" arose from the idea of a measured or proportioned object serving as a pattern for something larger. In mathematics, modeling means constructing a proportional and structured representation based on fundamental assumptions (*axioms*) together with defined relationships (*equations*) in order to reason about and predict how a system behaves.

The story of mathematics may begin in the realm of myth, where ancient narratives tell of gods and celestial beings who bestowed advanced knowledge upon humanity. These accounts can be read as symbolic records of the sudden appearance — or reappearance — of sophisticated mathematics and astronomy in early societies.

In Mesopotamia, the god Enki was said to have given humankind the ability to measure, a skill that underpinned agriculture, architecture, and trade. In Egypt, Thoth was revered as the god of wisdom, writing, and mathematics, a role that emphasized the importance of numerical knowledge in maintaining harmony between the human and cosmic orders. Such myths present mathematics not merely as a practical invention, but as something inherent to the structure of the universe, a language capable of linking earthly life to higher patterns.

Some modern interpretations, like those advanced by Graham Hancock, suggest that these myths may preserve memories of a highly advanced prehistoric civilization — a "lost" culture with mastery of astronomical and mathematical principles far beyond what is usually imagined for its time. The architectural precision of the pyramids of Giza, aligned with stellar patterns, and the vast geometric figures of the Nazca Lines, visible only from the air, are among the examples often cited as signs of such knowledge.

Monumental sites such as Göbekli Tepe, Stonehenge, and the megalithic temples of Malta show that prehistoric builders possessed a profound grasp of geometry, astronomy, and engineering. The apparent alignments of these structures with solstices and equinoxes suggest a deliberate connection to celestial cycles, implying both sustained observation and the ability to translate those patterns into large-scale architectural design.

Whether such skills were understood as gifts from the divine, as in the ancient myths, or as the product of human ingenuity and necessity, mathematics has been deeply embedded in human development since its earliest recorded traces. It served not only as a tool for solving practical problems, but as a means of connecting human activity to the perceived order of the cosmos.

The Sumerians of Mesopotamia developed one of the earliest known writing systems, *cuneiform*, first used to record trade agreements, inventories, and administrative decrees, and later extended to capture mathematical knowledge. By around 2500 BCE, Sumerian mathematics was operating within a base-sixty numeral system (*sexagesimal*) — a structure that still echoes today in the three hundred sixty degrees of a circle and the sixty minutes in an hour. The choice of sixty was likely shaped by both astronomical cycles and practical needs: the year's passage across the heavens could be broken into useful segments for calendars, agriculture, and governance, and the number's many divisors made it ideal for precise fractional calculations. Clay tablets from this era reveal work on formal equations, geometric reasoning, and methods for calculating areas and volumes. This literacy in numbers was not an abstract luxury — it was a core technology for managing city-states, organizing labor forces, dividing farmland, and tracking goods across long-distance trade routes.

When ancient surveyors and builders faced practical measurement problems, they often reduced them to idealized forms — shapes such as squares, rectangles, or cubes that yielded clear, exact relationships. In these models, the logic of equations emerged naturally from the physical task. Measuring the length of a rectangular plot, for instance, produced a linear relationship: each unit increase in length caused a constant, proportional increase in area. But if the challenge was to determine the side length of a square field from its total area, the relationship became *quadratic*: the area is the product of a side multiplied by itself, and solving for the side requires undoing that squared relationship. The same logic applied to *cubic* problems, such as determining the edge length of a grain storehouse from its volume — the volume being the side multiplied by itself three times.

Though the physical world often presented irregular shapes and imperfect conditions, ancient mathematicians worked with these ideal forms as conceptual training grounds. Their unambiguous, single-solution nature made them excellent tools for teaching and for developing methods of reasoning. By practicing on perfect squares and cubes, they honed techniques that could later be adapted to the messier demands of real construction, taxation, and trade — preserving a chain of logic that connected pure numerical thought to the tangible needs of their world.

Following the decline of the Sumerian city-states around 1894 BCE, the Babylonians rose to prominence and inherited the mathematical traditions of their predecessors. The kinds of relationships seen in Sumerian problem-solving — whether in the measurement of land, the determination of dimensions from area or volume, or other proportional calculations — were carried forward with greater refinement and systematization. Babylonian scribes organized these methods into structured procedures and applied them across a wide range of contexts, from surveying and architectural planning to calculating interest and distributing resources. Their most demanding applications were in astronomy, where the long-term tracking of celestial movements required exceptional precision. In this setting, mathematics became a predictive science, capable of charting planetary and stellar positions years in advance and integrating those patterns into both administrative and ceremonial calendars.

Two of the most important surviving records of ancient Egyptian mathematics are the *Moscow Mathematical Papyrus*, dated to around 1850 BCE, and the *Rhind Mathematical Papyrus*, dated to around 1550 BCE. The Moscow Papyrus contains twenty-five worked problems, including geometric procedures such as calculating the volume of a truncated pyramid. The Rhind Papyrus is broader in scope, with examples in arithmetic, formal equations expressed in a narrative style, and a sophisticated treatment of fractions. These texts show that Egyptian scribes worked with linear problems and proportional reasoning, and were able to handle complex numerical relationships with precision.

Egyptian mathematics was expressed in a rhetorical form, where each calculation was written out as a sequence of steps in words rather than symbols. A problem in the Moscow Papyrus for finding the volume of a truncated pyramid (frustum) gives the method in explicit instructions:

> If you are told: a truncated pyramid of 6 for the vertical height by 4 on the base by 2 on the top: You are to square the 4; result 16. You are to double 4; result 8. You are to square this 2; result 4. You are to add the 16 and the 8 and the 4; result 28. You are to take 1/3 of 6; result 2. You are to take 28 twice; result 56. See, it is of 56. You will find [it] right.
>
> — Gunn & Peet, *Journal of Egyptian Archaeology* (1929)

This is the same sequence of operations that, in modern symbolic form, expresses the frustum volume relationship, but in the Egyptian presentation it appears as a verbal algorithm. The clarity of the steps ensured that the method could be followed and verified without reference to abstract notation.

Mathematics in Egypt was closely tied to the organization of daily life. *Aha* problems solved for unknown quantities, often in the context of allocating rations or other resources. *Pefsu* problems calculated ratios for producing bread and beer, ensuring consistent quality and efficient use of ingredients. Both required proportional reasoning and a form of equation-solving to address practical challenges in food production and resource management.

Other categories reflected administrative and economic needs. *Baku* problems functioned as audits, checking tallies of state and temple inventories against expected values. *Ship's Part* problems dealt with fair division of goods or wages among a group, ensuring proportional shares for crews, labor teams, or other collective efforts. Geometric calculations, including those for the volume of a frustum, were essential for architectural design and construction on a monumental scale.

Civilizations such as the Minoans and Mycenaeans, flourishing in the Aegean during the Late Bronze Age (circa 1550-1200 BCE), absorbed and built upon Egyptian and Mesopotamian administrative and mathematical practices. Their record-keeping, early writing systems, and architectural feats reveal that the principles of order, measurement, and even rudimentary logical reasoning were continually evolving. Additionally, trade networks and the influence of Phoenician merchants helped disseminate ideas across the Mediterranean, ensuring that the intellectual legacy of Egypt and Mesopotamia was neither lost nor stagnant but rather transformed in new and subtle ways.

After the decline of Egypt's New Kingdom (circa 1550–1070 BCE) and the tumult of the Late Bronze Age collapse (circa 1200 BCE), Greece entered what became known as the Greek Dark Age (circa 1100–800 BCE). Although written records from this period are scarce, archaeological sites such as Lefkandi demonstrate that the seeds of social organization, trade, and technical skill persisted. By 800 BCE, with the reintroduction of writing via the Phoenician alphabet, the stage was set for a cultural rebirth during the Archaic period (800–600 BCE), when Greek society began to transform mythic narratives into frameworks for understanding the world.

By around 600 BCE, mathematical thought in the Mediterranean was increasingly associated with Greek philosophers such as Thales and Pythagoras, who explored ways of treating number and form as subjects of reasoning independent from immediate practical application. This shift toward abstraction laid the groundwork for a tradition in which mathematics could be developed, analyzed, and debated as a discipline in its own right.

At the dawn of Greek philosophy, Thales (circa 624–546 BCE) sought natural, rational explanations for phenomena traditionally explained through myth. He is credited with proposing that water was the fundamental substance underlying all things — an idea recorded by later authors such as Herodotus (circa 484-425 BCE) and Aristotle — which marked a shift toward viewing the world as governed by principles that could be understood and reasoned about.

Thales is also associated with several geometric propositions, among them the theorem that bears his name: in any circle, a triangle formed by the diameter and any third point on the circumference is a right triangle. This can be seen by drawing a radius from the third point to the circle's center, creating two isosceles triangles whose angle relationships require the remaining angle to be a right angle, since it is the sum of the other two angles. The practical implication of this theorem is that one can easily construct perfect right triangles using only a compass and a straight edge.

Thales, equipped with his mastery of triangles, is said to have demonstrated how to measure the heights of pyramids and the distances of ships from shore. These feats are some of the earliest recorded examples of applying mathematical reasoning to solve real-world problems. Herodotus also wrote that Thales predicted a solar eclipse in 585 BCE, an event so striking that it reportedly halted a battle between the Lydians and the Medes, further enhancing his reputation for insights with tangible consequences.

Pythagoras (circa 570–495 BCE), a near contemporary, advanced the rational study of nature through the lens of number and proportion. The Pythagorean school, as later described by Iamblichus (circa 245-325 CE) and Porphyry (circa 234-305 CE), blended disciplined mathematical investigation with symbolic and mythic elements, using both to convey abstract ideas. The Pythagoreans are most widely associated with the geometric relationship in right triangles in which the sum of the squares of the two shorter sides equals the square of the hypotenuse. Beyond this, they held that number itself was the underlying essence of reality, binding together the structure of the cosmos, the patterns of music, and the order of moral life.

Pythagoras's wife Theano (circa 560-470 BCE), a philosopher in her own right, clarified this principle after his death:

> I have learned that many of the Greeks suppose Pythagoras said that everything came to be from number. This statement, however, poses a difficulty — how something that does not even exist is thought to beget things. But he did not say that things came to be from number, but according to number.
>
> — *On Piety* translated by Walter Burkert, *Lore and Science in Ancient Pythagoreanism* (1972).

The Pythagoreans envisioned a kind of mathematical atomism in which all quantities could be expressed as whole units or sums of such units. Yet in attempting to find a smallest unit that could measure both the side and the diagonal of a square, using a technique known as *infinite descent*, they encountered a profound difficulty. By reasoning that any candidate unit could always be halved — and halved again without limit — they arrived at the unsettling conclusion that no indivisible unit existed for certain geometric ratios, such as that between a square's side and its diagonal. This was the problem of *incommensurability*, and for a philosophy rooted in finite, discrete measures, it posed a direct challenge to the coherence of their worldview. The discovery was guarded within the school, and legend holds that a member who revealed it to outsiders was drowned at sea, a grim reflection of how deeply this contradiction shook the foundations of their thought.

Contemporaneous with the Pythagorean school, Heraclitus of Ephesus (circa 535–475 BCE) offered a vision of the cosmos in which change was fundamental. His doctrine "everything flows" (Πάντα ῥεῖ, panta rei) expressed the view that all things are in constant transformation. Yet within this perpetual flux, he proposed the existence of the *logos* (λόγος) — a unifying rational principle that governs and orders change. The surviving fragments of his thought, preserved by later writers such as Plato (circa 428-348 BC) and Diogenes Laertius (circa 180-240 CE), combine mythic imagery with a drive toward identifying an intelligible structure underlying the shifting appearances of the world.

Parmenides of Elea (circa 515–450 BCE), in his poem *On Nature*, advanced a sharply contrasting view. He argued that reality is a single, unchanging, indivisible whole, encapsulated in the dictum: "what is, is, and what is not, is not". For Parmenides, the senses mislead by presenting a world of change and multiplicity, while rational thought reveals a permanent and undivided being. His emphasis on logical necessity over sensory evidence became a defining influence on later Greek philosophy.

Parmenides's student Zeno of Elea (circa 490–430 BCE) developed a series of paradoxes intended to expose contradictions in the common conception of motion and plurality. In the *Achilles and the Tortoise*, he reasoned that if space and time are infinitely divisible, a faster runner could never overtake a slower one, for the pursuer must first reach an infinite sequence of intermediate points. In the *Dichotomy*, motion itself appears impossible because one must first traverse half a given distance, then half of the remainder, and so on without end. The *Arrow* argued that at any single instant of time, a moving arrow is at rest, and if time is composed entirely of such instants, motion cannot occur. These paradoxes confronted philosophers and mathematicians with profound questions about continuity, infinity, and the nature of space and time — challenges that would echo through subsequent centuries of thought.

Philolaus (circa 470–365 BCE) developed the idea of a cosmic "harmony of the spheres", proposing that the motions of the heavenly bodies correspond to musical relationships governed by number. In this view, mathematical ratios order the cosmos just as they order consonant intervals in music. Philolaus also described a cosmology centered on a "central fire", with Earth and the other bodies revolving around it, an arrangement intended to express numerical balance and cosmic symmetry rather than to mirror everyday appearances of the sky.

Hippocrates of Chios (circa 470–410 BCE) advanced geometry toward a more systematic, deductive practice. Although his writings do not survive intact, later commentators such as Simplicius (circa 490–560 CE) and Proclus (circa 412–485 CE) report that his work, including a collection sometimes referred to as *Elements of Geometry*, helped shape the tradition seen in later compilations with similar names. Hippocrates is especially associated with the quadrature of *lunes* — crescent-shaped regions bounded by arcs. His method compared areas of circular segments to areas of straight-edged figures through carefully chosen constructions and proportional relations. The result showed that certain curved regions admit exact area determinations by pure reasoning, establishing a pathway for treating curved boundaries with the same logical precision long applied to rectilinear figures.

Democritus (circa 460–370 BCE), along with his mentor Leucippus (circa 480-420 BCE), proposed that the universe is composed of an infinite number of indivisible particles — *atomos* (ἄτομος) — that move through the void. Although Democritus did not use the language of modern mathematics, his view was that apparent continuity is an illusion; the smoothness of a line or the even distribution of mass in a sphere is not due to the infinite divisibility of matter, but rather the cumulative effect of countless, finite, indivisible parts. In his view, although atoms are not *infinitesimal* in the rigorous sense later defined in calculus, they represent the irreducible building blocks from which continuous change is constructed. Although many of his mathematical writings have not survived, historical references suggest that he engaged with problems involving proportion, measurement, and the nature of the continuum.

In the 5th century BCE, Meton of Athens (circa 470-400 BCE) integrated systematic astronomical observation with mathematical reasoning in the formulation of the *Metonic cycle*. This cycle demonstrated that two-hundred-thirty-five lunar months closely match nineteen solar years, differing by only about two hours per cycle. The model addressed a practical and recurring problem: synchronizing lunar and solar calendars for religious, agricultural, and civic purposes. Euctemon (circa 490-420 BCE) expanded its application by embedding the cycle into functional calendrical systems, improving the coordination of civic and ritual life. Callippus (circa 370-300 BCE) later refined the framework, introducing adjustments to reduce residual discrepancies and yielding a more accurate and durable basis for timekeeping.

During the same century, the Sophists fostered a climate of critical inquiry, questioning traditional assumptions and valuing the power of rhetoric and structured argument. In this environment, Antiphon (circa 480-411 BCE) considered the longstanding challenge of determining the area within a circle. Accurate land measurement had been essential in earlier civilizations for taxation, resource allocation, and infrastructure planning, particularly in contexts like Egypt where annual Nile floods required the resurveying of farmland. Surveyors could readily determine areas bounded by straight lines by subdividing them into triangles or rectangles, but circular boundaries resisted exact calculation. Reservoirs, canals, or round plots surrounding temples and other important sites demanded a reliable method for deriving area from a single dimension, such as diameter.

Egyptian surveyors, as seen in the *Rhind Mathematical Papyrus*, estimated a circle's area by taking eight-ninths of its diameter as the side of a square. In effect, they were saying that a circle could be treated as nearly equal in area to such a square. This procedure produced an effective approximation of π for everyday purposes. Babylonian methods were likewise empirical, relying on fixed numerical values for π that sufficed for practical needs. Fortunately for mathematics, the Greeks were not solely focused on immediate practical utility; they were deeply invested in developing rigorous, systematic methods to understand the underlying principles of geometry. These methods would allow calculating the true area of a circle as closely as desired, and eventually led to techniques that would underpin calculus.

Antiphon introduced a systematic approach: by inscribing a polygon, almost certainly starting with a hexagon due to its natural construction from the circle's radius, and then doubling the number of sides, he could progressively refine the approximation of the circle's area. As the number of sides increased, the polygon seemed to approach the shape and area of the circle — although total proof of its accuracy would not be obtained until Eudoxus (circa 390-340 BCE).

In the generation following the Sophists, Plato, most notably in the *Timaeus* and *Critias*, set out his view of a rational principle shaping the cosmos. In this cosmology, a divine craftsman — the *Demiurge* — imposes order on pre-existing chaotic matter according to eternal forms, constructing the world through mathematical harmonies and the geometry of the regular solids. The structure of the universe, in Plato's account, is therefore not only rational but intrinsically mathematical. Alongside this vision, he introduced the tale of Atlantis, framing it as ancient history preserved through Egyptian records. Whether history or allegory, its role was to contrast a wealthy but corrupted civilization with an idealized Athens, reinforcing the parallel that just as the cosmos is sustained by rational and mathematical order, so too must human societies endure by grounding themselves in virtue.

Archytas of Tarentum (circa 428–347 BCE), a contemporary and associate of Plato as well as a follower of the Pythagorean tradition, was held in high esteem by Plato, as seen in the *Seventh Letter*. There, Plato praised Archytas's virtue, wisdom, and practical leadership. Archytas served multiple terms as *strategos* (στρατηγός, i.e. general), of Tarentum and has been regarded by some as a potential model for Plato's philosopher-king. His achievements encompassed both theoretical and applied fields, from mathematics and music to mechanics. Ancient sources even credit him with designing a self-moving wooden pigeon, possibly powered by a counterweight or steam.

Archytas's most renowned mathematical contribution was his solution to the Delian problem — the challenge of constructing a cube with double the volume of a given cube. The problem was said to originate from an oracle's command to the people of Delos to double the size of Apollo's cubical altar to end a plague. Misunderstanding the command, they doubled each side length, producing a cube eight times the original volume. The real task required doubling the volume, a problem that drew considerable attention from Greek geometers. Hippocrates of Chios had shown that the challenge could be reduced to finding two *mean proportionals* between a given length and twice that length. Building on this reduction, Archytas devised a three-dimensional solution: by analyzing the intersection of a cone, a cylinder, and a torus, he constructed the lengths that satisfied the mean proportional condition. This provided a rigorous theoretical answer to the problem, though it lay outside the limitations of constructions using only compass and straightedge — a point on which Plato expressed dissatisfaction.

Archytas extended his studies into *harmoniai* (ἁρμονίαι, i.e. musical harmony). While the term *frequency* did not exist in his vocabulary, he understood pitch in terms of the numerical ratios of lengths. Likely working with monochords or similar instruments, he observed that consonant musical intervals arose from simple ratios between string or pipe lengths. Later accounts describe his articulation of these relationships in terms of *arithmoi* (ἀριθμοί, i.e. numbers) and *logoi* (λόγοι, i.e. ratios). He identified the *diapason* (διαπασῶν, i.e. octave) with a 2 : 1 ratio, the *diapente* (διὰ πέντε, i.e. fifth) with 3 : 2, and the *diatessaron* (διὰ τεσσάρων, i.e. fourth) with 4 : 3. In the Pythagorean tradition, these proportions were more than musical facts; they were expressions of a universal *harmonia* (ἁρμονία, i.e. harmony), a mathematical order underlying both the structure of the cosmos and the consonance of sound.

Theaetetus (circa 417–369 BCE), a friend of both Socrates and Plato, appears as the main interlocutor in Plato's dialogue *Theaetetus*. The work is framed as an extended inquiry into the nature of knowledge, with Socrates posing the central question and Theaetetus offering successive definitions. Theaetetus first suggests that knowledge is equivalent to perception, a position resembling *subjective relativism*; later, he considers whether it might be true judgment, and finally, whether it could be true judgment accompanied by an explanatory account (*logos*). Socrates examines each proposal in turn, uncovering difficulties through counterexamples and logical analysis. If knowledge were merely perception, the variability of sensory experience would make it unstable; if it were simply true judgment, then without justification it would lack completeness. The dialogue ends without a final answer, leaving the question unresolved in a manner characteristic of many Platonic dialogues, but in doing so it illustrates the complexity of defining knowledge and the necessity of continued investigation. Theaetetus's tentative proposals and Socrates's method of dialectical questioning reveal that knowledge is far more complex than any single, straightforward definition can capture, suggesting that further inquiry is needed to fully understand this elusive concept.

Theaetetus's mathematical contributions were substantial in two distinct areas. The first addressed the longstanding problem of incommensurable magnitudes, known at least since the Pythagoreans, in which certain lengths — such as the diagonal of a square — cannot be expressed as a ratio of whole numbers. Seeking to integrate these quantities into a coherent geometric framework, Theaetetus classified different types of irrational magnitudes and developed a systematic method for treating them within geometry. His work created an ordered structure for handling incommensurables and prepared the ground for later refinements by mathematicians such as Eudoxus.

His second major contribution was the determination of all possible *regular convex polyhedra*. A regular convex polyhedron has identical faces that are each *regular polygons* (equilateral shapes), with the same number of faces meeting at each vertex, and no points protruding beyond the convex boundary. This uniformity ensures that any two points on its surface can be connected by a straight line lying entirely on or within the solid. For such a solid to exist, the flat shapes connecting at each corner must leave a gap in the space around the vertex; this gap allows the shapes to tip inward toward each other and form the surface of a solid that bulges outward, rather than lying flat in a plane.

Theaetetus explored which regular polygons could meet these conditions. If three regular triangles are joined edge to edge around a single point, there is a noticeable gap, allowing them to fold up into a three-sided pyramid, the *tetrahedron*. Four triangles joined in this way can fold into the *octahedron*, and five triangles can form the *icosahedron*. Three squares meeting at a point also leave a gap that allows folding into a cube, but if a fourth square is added, the shapes close together into a flat arrangement with no room to fold. Three regular pentagons meeting at a point can be folded into the *dodecahedron*, but any attempt to use regular hexagons or shapes with more sides results in their edges lying flat or overlapping, which makes it impossible to build a bulging, fully enclosed solid.

By working through each possibility in turn, Theaetetus showed that only five such solids could be made: the tetrahedron, cube, octahedron, dodecahedron, and icosahedron. These shapes were later called the *Platonic solids* and were given symbolic associations with the elements and the structure of the cosmos. His classification, achieved without the formal system that Euclid (circa 325-265 BCE) would later supply, reflects a remarkable level of geometric insight built up from direct reasoning about space and form.

Eudoxus of Cnidus (circa 408–355 BCE), a Greek mathematician and student of Plato, extended the work of Theaetetus by finding a way to handle a problem that had unsettled Greek mathematics for generations: how to compare lengths that could not be measured exactly by the same unit. None of his writings survive, and what is known of his work comes largely from later authors, most notably Euclid.

Earlier definitions of ratio assumed both quantities could be broken down into a shared unit of measure. This worked when both were whole-number multiples of some smaller length, but it failed when no such common unit existed — as in the case of a square's side and its diagonal. In those cases, the ratio could not be expressed as a fraction of whole numbers, and so it lay outside the formal tools available for proofs.

Eudoxus solved this by redefining what it meant for two ratios to be equal. His idea did not depend on finding a common measuring unit at all. Instead, he said: take any whole-number multiples of the first two quantities in one ratio, and take the same multiples of the two quantities in the other ratio. If in every such comparison the results are always in the same relationship — both greater, both equal, or both less — then the two ratios are equal. This definition worked in all situations: when a common unit existed, it agreed with the older definition; when no common unit existed, it still gave a consistent, reliable way to compare magnitudes. By doing this, Eudoxus allowed ratios involving *incommensurable* quantities to be used alongside ordinary ones in geometric reasoning, removing a major obstacle to the development of proofs.

Eudoxus is also remembered for perfecting a method of finding the exact area or volume of shapes bounded by curves — something that could not be measured directly with straight-edged constructions. The idea was to trap the curved figure between shapes whose areas or volumes could be calculated exactly, and to keep narrowing the gap until nothing measurable remained.

An earlier version of this approach had been suggested by Antiphon, who proposed drawing a polygon inside a circle and increasing the number of its sides step by step so it fit more and more closely. This gave better and better lower estimates for the circle's area, but it still left open the possibility that the circle's true area was larger by some amount.

Eudoxus's refinement was to work from both sides at once. He drew another series of polygons outside the circle, each slightly larger than the circle. As the number of sides increased for both the inner and outer polygons, the areas of the two shapes grew closer together. The circle's area was 'trapped' between them, and because the gap between inner and outer shapes could be made as small as desired, the area of the circle could be identified exactly as the single value caught between the two.

This method — later called the *method of exhaustion* — turned what had been a plausible estimation into a rigorous proof technique. It allowed Greek mathematicians to work out the exact areas of circles, the volumes of spheres, and the shapes of other curved solids with the same logical certainty as they handled polygons and polyhedra.

Among the celebrated challenges of Greek geometry, *squaring the circle* (constructing a square with the same area as a given circle) stood alongside cube doubling (the Delian problem) and *angle trisection* (cutting an angle in three equal parts) as tests of skill and method. These problems were to be solved using only a straightedge and a compass, the basic tools of a geometer. They were accepted not simply by tradition, but because the Greeks could clearly take as given that each carried out its operation: a straightedge could be used to draw a straight segment between two points, and a compass to draw a circle from a given center through a given point. Any construction using them could be broken down into a sequence of these simple, unquestioned steps.

Hippias of Elis (circa 460–399 BCE), in seeking a means of dividing an angle into an exact number of equal parts, defined a curve called the *quadratrix*, produced by combining two uniform motions: one line sweeping out a quarter circle by pivoting through a right angle at a steady rate, and another moving straight down at the same steady rate along the side of a square. Because the motions were perfectly synchronized, the curve linked each angular position of the rotating line to a precise point along the square's side, making it possible to mark proportional divisions of the angle by marking proportional divisions along the side and connecting them through the curve.

Dinostratus (circa 390–320 BCE) applied this curve to the problem of squaring the circle, showing how, if one used the quadratrix, the side of the required square could be determined exactly. The reasoning was correct, but some contemporary geometers objected to it. The difficulty was not in the reasoning but in the premise: the construction assumed that one could already produce the quadratrix itself. To do so required a device capable of coordinating two separate uniform motions in perfect synchrony. To accept the quadratrix as given was to introduce a curve whose existence was unproven, resting the solution on a premise that could not be justified within the rules of the art. This is why, despite its ingenuity, the method was regarded as lying outside the accepted bounds of 'pure' geometry.

Menaechmus (circa 380–320 BCE), likely a pupil of Eudoxus of Cnidus, is remembered for discovering a new family of curves — the *conic sections* — while seeking a solution to the Delian problem (the challenge of constructing a cube with twice the volume of a given cube). Archytas had produced a solution a generation or two earlier, but his approach used a three-dimensional construction involving the intersection of a cylinder, cone, and torus, a method admired for its ingenuity but difficult to visualize or reproduce in the two-dimensional diagrams of standard geometry. The search continued for a simpler, more 'static' construction that could be carried out in the plane.

Menaechmus's insight was that certain curves formed by slicing a cone with a plane at specific angles contained within their proportions the relationships needed for the problem. He produced two such curves: one in which the square of a point's distance from a fixed line was proportional to its distance from a fixed point (later called a *parabola*), and another in which the product of two distances was constant (later called a *hyperbola*). By analyzing the similar triangles built into these constructions, he showed that by cutting cones at just the right inclinations, each of these relationships would be satisfied. When two such conic sections were drawn together in a particular arrangement, their intersection point provided two lengths in mean proportion between a given length and its double — exactly the condition required to double a cube's volume. As with the quadratrix, the result was mathematically sound, but the curves themselves were not obtainable by straightedge-and-compass construction, placing the method outside the strict canon of pure geometry.

Euclid's *Elements* (circa 300 BCE) epitomized that canon. In this monumental compilation, Euclid organized the accumulated geometry of his time into thirteen books built from a small foundation: a set of definitions, five postulates, and a handful of common notions. From these starting points, he deduced hundreds of propositions through rigorous, step-by-step proofs. The work formalized and extended results from Eudoxus, Hippocrates, Theaetetus, and others, while also preserving practical geometric knowledge inherited from earlier cultures such as the Egyptians and Babylonians. It drew upon previous compilations, like the work of Theudius of Magnesia (circa 375–315 BCE), who had earlier compiled a systematic set of theorems and made many partial results more general. Its deductive method, in which each result followed logically from what had been established before, became a model for mathematical reasoning itself. Surviving through medieval manuscript traditions — particularly in the Byzantine world — the Elements shaped both mathematics and logical thought for over two thousand years.

In contrast, Aristotle (384–322 BCE), another student of Plato and later tutor to Alexander the Great (356–323 BCE), developed a framework for more general purpose reasoning. In his collection of works known as the *Organon*, he set out the theory of *syllogistic logic* — a formal structure for deductive argument in which two premises combine to yield a conclusion. A classic example is: "All humans are mortal; Socrates is a human; therefore Socrates is mortal".

Aristotle's concern with valid inference extended into all areas of his work. His writings on metaphysics, ethics, natural science, and rhetoric shaped the intellectual tradition that followed him for centuries. In natural science, he united systematic classification with direct observation, insisting that knowledge should be ordered and grounded in what could be perceived. This empirical habit, combined with his logical rigor, helped establish principles that would later inform the *scientific method*.

In works such as *Physics* and *Metaphysics*, Aristotle also addressed fundamental questions about infinity and continuity, sharply distinguishing between what he called "potential infinity" and "actual infinity". A magnitude — such as a line segment or an interval of time — could in principle be divided without limit, but this was only a potential infinity, since the divisions were never all completed. From this standpoint he responded to Zeno of Elea's paradoxes, which argued that motion was impossible because it would require completing infinitely many steps. For Aristotle, the steps existed only potentially, so motion could proceed without the need to finish an actual infinite sequence.

The Stoic school, founded in Athens by Zeno of Citium (circa 334–262 BCE), shared with Aristotle and Euclid the conviction that the world is intelligible and that human reason can grasp its order. In Stoic thought, that order was the *logos* (λόγος), a Greek term meaning 'word', 'account', or 'reason'. The logos was the active, rational principle structuring the cosmos, the same principle that enables human beings to think and to reason. While Aristotle examined the forms of valid argument and Euclid codified the rules of geometric proof, the Stoics asked why such reasoning is possible at all. Their answer was that the rational structure of the universe and the rational faculty in humans are two aspects of the same underlying order.

For the Stoics, logic was one branch of a unified philosophy, alongside physics and ethics. Studying logic trained the mind to follow the logos; studying nature revealed the scope of the logos in the world; practicing virtue meant living in harmony with it. In this way, their focus on reason was inseparable from their view of how the cosmos works and how life should be lived. Although most original Stoic writings are lost, later sources describe a system that treated the laws of thought and the laws of nature as expressions of a single, coherent order. The idea of the logos would remain influential long after the Stoic school itself had faded, appearing in early Christian writings such as the *Gospel of John*, where it was used to describe Jesus as the embodiment of divine reason and order — a formulation that would shape much of medieval European thought.

In the cosmology of the Stoics, Plato, and Aristotle, the universe was finite, spherical, and enclosed by a sphere of fixed stars, with the Earth at its exact center. This geocentric arrangement was justified from what they saw as rational and empirical grounds. The heaviest elements, earth and water, were thought to move naturally toward the center of the universe, and the Earth's stability at that point explained, for them, why objects dropped from a height fell straight down rather than being left behind. They reasoned that if the Earth were moving rapidly through space, the stars would appear to shift against one another over the course of a year — a phenomenon called *stellar parallax* — yet no such shift was visible to the naked eye. Likewise, they believed that if the Earth were in motion, constant high-speed winds would sweep over its surface; the fact that no such universal wind was observed was taken as further confirmation that the Earth was stationary. For the Stoics, a central, motionless Earth fit both their physical theory and their view of a purposeful cosmic design.

Motivated by a desire to provide a more coherent and mathematically consistent model of the universe than existed at the time, Aristarchus of Samos (circa 310-230 BCE) proposed a model in which the Earth was just one of several planets orbiting the Sun. This was a radical departure from the dominant theories of Plato and Aristotle, and was rejected even generations later by Ptolemy (circa 100-170 CE). Aristarchus observed the shadow cast on the Moon by the Earth during a lunar eclipse, and from its curvature estimated that the Earth was about twice the size of the Moon. He also reasoned that when the Moon appears half-illuminated as seen from Earth — meaning that one-quarter of its total surface is visible — the Sun, Earth, and Moon should form a right triangle, with the right angle at the Moon. By eyeballing the angular distance between the Moon, himself, and the Sun, he judged the angle at his position to be slightly less than a right angle, and thus deduced that the Sun must be much farther away than the Moon. Given the greater distance of the Sun yet its seemingly similar size in the sky, he concluded that the Sun must be significantly larger than both the Moon and Earth. Taken together, these observations suggested to Aristarchus that a model of the cosmos with the Sun at the center would be most logical.

An admirer of Aristarchus, Archimedes of Syracuse (circa 287–212 BCE) is widely regarded as one of antiquity's greatest mathematicians and engineers. Ancient sources credit him with the discovery of the principle of buoyancy, famously associated with his exclamation "Eureka!". According to later accounts, he realized while bathing that a submerged object experiences an upward force equal to the weight of the fluid it displaces — an insight that provided a way to determine the purity of gold without damaging it and established a fundamental law of fluid mechanics. One popular version of the story holds that King Hiero II of Syracuse suspected a goldsmith had adulterated a crown with a less dense metal. To test this, Archimedes is said to have immersed the crown in water and measured the displacement, then compared this with the displacement from an equal weight of pure gold.

Archimedes was reputed to be so absorbed in his studies that he became oblivious to his surroundings, even in times of crisis. Plutarch (circa 46-120 CE) relates that during the Roman siege of Syracuse in 212 BCE, Archimedes was so engrossed in drawing geometric figures in the sand that he ignored the commotion around him. A Roman soldier approached, demanding his attention, and Archimedes uttered something to the effect of "Do not disturb my circles" before being killed on the spot.

In mechanics, Archimedes investigated the properties of levers, demonstrating that a relatively small force, applied at a greater distance from a pivot, could balance a much larger weight — a principle that would influence the design of machinery and military devices. He also devised the Archimedean screw, a helical surface wrapped around a cylinder, which could lift water when rotated. This device was used for irrigation and drainage, showing how mathematical insight could yield practical engineering solutions.

Archimedes is credited with authoring around fifteen treatises, including *On the Equilibrium of Planes*, *Measurement of a Circle*, and *On the Sphere and Cylinder*. These works, written in Greek and circulated among scholars, were composed over his mature years and reveal a mind capable of transforming simple observations into general mathematical laws.

Building on the method of exhaustion developed earlier by Eudoxus and Antiphon, Archimedes applied it with unmatched rigor to determine exact areas and volumes. He used it to compute the area of a circle, and the volume and surface area of a sphere, with a precision that astonished his contemporaries. In *On the Sphere and Cylinder*, he established that the volume of a sphere is two-thirds the volume of the smallest cylinder that encloses it — a result he regarded as his greatest achievement. He is said to have requested that a sphere inscribed in a cylinder be engraved on his tombstone.

Archimedes's proof was characteristically ingenious. Consider a sphere enclosed in a cylinder of the same diameter and height. If two cones are placed tip-to-tip within the cylinder, each base equal to the sphere's diameter and each base touching one end of the cylinder, then at any height, the combined cross-sectional area of the sphere and the cones equals the cross-sectional area of the cylinder. Since the volume of a cone is known to be one-third that of a cylinder with the same base and height, the volume of the sphere can be deduced as the cylinder's volume minus the cones' combined volume. By comparing corresponding slices rather than summing infinitely many small elements directly, Archimedes transformed a seemingly intractable problem into a triumph of geometric reasoning.

Archimedes described such insights in *The Method of Mechanical Theorems* (Μέθοδος τῶν μηχανικῶν θεωρημάτων), a treatise he addressed to Eratosthenes (circa 276–194 BCE), the chief librarian at the Library of Alexandria. In it, Archimedes explained how he discovered new results by mentally dividing figures into infinitely many thin slices and using the mechanical analogy of a lever to balance slices of an unknown figure against those of a known one. This balancing revealed proportional relationships that could then be proved rigorously by the method of exhaustion.

Born in the Greek colony of Cyrene (in modern Libya), Eratosthenes had access at Alexandria to an unparalleled collection of manuscripts. His own contributions were wide-ranging. He devised the systematic *sieve of Eratosthenes* for identifying prime numbers by progressively eliminating composite integers. He also applied geometry to the measurement of the Earth. Reports from Syene (modern Aswan, Egypt) told that at noon on the summer solstice, the sun shone directly to the bottom of a vertical well. On that date in Alexandria, Eratosthenes measured the angle of a gnomon's shadow as about one-fiftieth of a full circle. Assuming Syene lay due south of Alexandria — a close but imperfect approximation — he reasoned that this angle was equal to the central angle between the two cities. If one-fiftieth of the circumference equaled the measured distance between them, then multiplying that distance by fifty yielded the Earth's circumference. His estimate may have been accurate to within about one percent, or at worst about fifteen percent, though it cannot be said for sure due to uncertainty about the length of the *stadion* unit he used.

Eratosthenes, with his pragmatic approach to mathematics, also decided to try his hand at the legendary Delian problem – the age-old challenge of doubling a cube's volume. Earlier attempts had relied on intricate curves or abstract constructions; Eratosthenes instead built the *mesolabium*, a frame holding a series of parallel plates that could slide to satisfy a specific proportional relationship. This mechanism could be used to find two mean proportionals between a given length and its double — in other words to take a cube root — thus providing a practical method to determine the side of a cube with twice the original volume.

Eratosthenes's career in Alexandria overlapped with that of another major figure of Hellenistic mathematics — Apollonius of Perga (circa 262–190 BCE), often described as "the great geometer". Apollonius lived primarily in Alexandria during the reigns of pharaohs Ptolemy III Euergetes and Ptolemy IV Philopator. While the surviving historical record offers few details of his life, his enduring reputation rests on his systematic treatment of curves in *Conics* (Κωνικά).

In the broader development of Greek mathematics, Apollonius served as a bridge between the intuitive geometric constructions of earlier mathematicians and the more abstract, analytic approaches that would appear centuries later. Figures such as Euclid and Menaechmus had provided essential groundwork in the study of conic sections, but Apollonius transformed this area into a coherent and general theory.

One of Apollonius's most significant contributions was to show that the ellipse, parabola, and hyperbola could all be generated from a single *double-napped cone* — two identical cones placed apex to apex — by varying the inclination of the cutting plane. Before this, each curve had generally been associated with a different type of cone: right-angled for the parabola, acute-angled for the ellipse, and obtuse-angled for the hyperbola. Apollonius demonstrated that a single cone sufficed: a plane parallel to the base produces a circle; a slight tilt yields an ellipse; a plane parallel to a generating line (a straight line from apex to base perimeter) produces a parabola; and a plane that cuts through both *nappes* forms a hyperbola. This unification eliminated the need for separate definitions and proofs for each case, revealing that all three curves were related manifestations of the same underlying geometry.

His originality lay not only in defining the conic sections as a unified family, but in formulating definitions that led to powerful consequences. Euclid had defined the diameter of a circle simply as any line passing through its center. Apollonius generalized this for all conic sections: a diameter is a straight line that bisects every chord drawn parallel to a given direction. This shifted the emphasis from a fixed point (the center, which may not be obvious for an ellipse or hyperbola) to a functional property of the figure. In practical terms, if parallel chords are drawn across an ellipse, the midpoints of all these chords will lie on a straight line — that line is the diameter. This approach made it possible to locate a diameter, and from it the center, even when the latter could not be assumed.

Apollonius also defined pairs of *conjugate* diameters in ellipses and hyperbolas: each diameter in the pair bisects chords parallel to the other. This property reveals a hidden symmetry of these curves, provides a reliable way to locate their centers (as the intersection of any two conjugate diameters), and underlies the geometric construction of tangents. These concepts gave mathematicians a deeper structural understanding of conic sections and laid groundwork for many later developments in geometry.

While Apollonius advanced the theory of pure geometry, Hipparchus of Nicaea (circa 190–120 BCE) was a pioneer in applying mathematics to the natural world. Often described as "the father of trigonometry", like Apollonius, Hipparchus worked in the Hellenistic period, an era shaped by the cultural exchanges made possible through the conquests of Alexander the Great. Drawing on earlier techniques such as the use of chord lengths to measure angles — a method inherited from Babylonian astronomy — he compiled one of the earliest known star catalogues, recording the positions and brightness of hundreds of stars with exceptional care. His sustained observations led him to identify the slow drift of the equinoxes relative to the fixed stars (i.e. the "precession of the equinoxes").

In the early Roman Empire, the intellectual emphasis turned increasingly toward practical applications of knowledge. Hero of Alexandria (circa 10 BCE – 70 CE) exemplified this trend. His *Metrica* (Μετρικά) served as a handbook of formulae for calculating areas and volumes, valuable to builders, architects, and landowners. Among these was Hero's formula for the area of a triangle, which required only the lengths of its sides — an elegant result that greatly simplified surveying and construction.

In contrast, Nicomachus of Gerasa (circa 60–120 CE) pursued a more philosophical form of mathematics. Drawing on the Pythagorean tradition, he concentrated on *number theory* rather than geometry or mechanics. His *Introduction to Arithmetic* (Ἀριθμητικὴ εἰσαγωγή, Arithmetike eisagoge) became a standard text for centuries in both the Islamic world and medieval Europe. In it, he classified numbers as odd, even, prime, composite, perfect, and more, often presenting these categories with the mystical and philosophical associations of Pythagorean thought. Unlike Euclid's *Elements*, Nicomachus's work avoided geometric proofs, instead using verbal explanations and numerical examples to express the intrinsic nature of numbers as fundamental constituents of reality.

For Nicomachus, the number one — the *monad* — represented unity, wholeness, and the ultimate source of all other numbers, often associated with the divine or the 'good'. The number two — the *dyad* — symbolized duality, opposition, and 'otherness', introducing multiplicity into the cosmos. These interpretations linked numerical properties directly to the structure and order of the universe. Nicomachus's synthesis of arithmetic with metaphysical symbolism deeply influenced later Neoplatonic philosophers and mathematicians, ensuring the survival of Pythagorean concepts well into the Middle Ages.

The longstanding aim of reconciling precise astronomical observation with the philosophical principle that the heavens move in perfect, uniform circles around the Earth led Claudius Ptolemy (circa 100–170 CE) to construct a model that both upheld this ideal and matched the data available to him. Working in Alexandria during the Roman period, and drawing on centuries of Greek, Babylonian, and Egyptian astronomical records, Ptolemy synthesized earlier ideas into a comprehensive geocentric system set out in his treatise *Almagest* (Μεγίστη Σύνταξις). Rejecting the heliocentric proposal of Aristarchus, he placed the Earth at the center and explained the motions of the Sun, Moon, and the five known planets — Mercury, Venus, Mars, Jupiter, and Saturn — through a combination of circular devices. Each planet moved on a small circle, the *epicycle*, whose center in turn moved along a larger circle, the *deferent*. To account for the apparent variations in planetary speed, particularly the retrograde motion, Ptolemy introduced the *equant*: an imaginary point offset from the deferent's center, from which motion along the deferent would appear uniform. By carefully adjusting the sizes and positions of these circles, he achieved a complicated model that appeased both the philosophical preference for uniform circular motion and the empirical record.

Ptolemy did not originate most of the components of his model, but he integrated them into a coherent, predictive framework that could serve navigation, calendar-making, and astrology. His work was part of the vibrant scholarly culture of Roman Alexandria, with access to rich libraries and earlier treatises. Though ultimately supplanted by heliocentric models using elliptical orbits, Ptolemy's system dominated astronomical thought for nearly fourteen centuries, standing as a pivotal fixture in mathematical astronomy.

Within the same Alexandrian tradition, Diophantus of Alexandria (circa 200–298 CE) emerged as a distinctive figure in mathematics. His precise dates remain uncertain, with scholarly estimates ranging from the 1st to the 4th century CE, though the 3rd century is most often accepted. He lived after the era of Euclid, Archimedes, and Apollonius, but still within an intellectually active environment. Diophantus's principal surviving work, *Arithmetica* (Ἀριθμητικά), is a collection of problems concerned with finding numerical solutions to algebraic equations.

Diophantus's *algebra* differed markedly from the symbolic form known today. He employed a mixture of words and abbreviations — a 'syncopated' notation — which represented an advance beyond the fully rhetorical style of his Greek predecessors, though still short of later symbolic systems. While Babylonians had developed methods for solving quadratic equations and earlier Greek mathematicians often recast such problems in geometric form, Diophantus approached them purely numerically, without recourse to geometric interpretation. He was also willing to accept fractional solutions, a departure from the Pythagorean focus on whole numbers.

The *Arithmetica* presents problems with worked examples, many of which fall into a category known as *Diophantine equations* — algebraic equations for which integer or rational solutions are sought. Diophantus treated linear and quadratic equations systematically and addressed certain cubic and higher-degree cases. His methods relied on substitutions and manipulations tailored to each problem, and he introduced a special symbol, resembling the Greek letter sigma, for the unknown quantity, along with abbreviations for its powers. Although he did not develop a general theory for all such equations — a challenge that remains unresolved today — his work marked a significant step toward an abstract, systematic treatment of algebra.

While other leading mathematicians of late antiquity in Alexandria continued to work within the inherited geometric tradition, Diophantus stood apart. Pappus of Alexandria (circa 290-350 CE) compiled the *Collection* (Συναγωγή), preserving and commenting on earlier works and occasionally solving problems with some overlap to Diophantus's interests, but always within a geometric framework. Eutocius of Ascalon (circa 480-540 CE) likewise devoted himself to commentaries on Archimedes and Apollonius, clarifying and transmitting classical geometry. Diophantus, by contrast, was not a commentator or a geometer in this tradition. His *Arithmetica* approached mathematics as a purely numerical discipline, independent of spatial construction, marking a decisive break from the Greek preference for geometry and laying essential groundwork for the development of algebra.

While Greek and Hellenistic traditions continued to shape mathematics in the Mediterranean world, sophisticated systems were also developing independently in other regions. In Mesoamerica, the Maya civilization (circa 2000 BCE – 1697 CE) created one of the most advanced mathematical frameworks of the ancient world. They used a *vigesimal* (base twenty) numeral system that included a positional symbol for zero, visible in inscriptions from around the 4th century CE. This was not merely an abstract development; it supported their complex calendrical and astronomical systems, which were essential for regulating agriculture and coordinating religious ceremonies.

In East Asia, Chinese mathematics developed independently, growing out of practical counting systems in use by around 1000 BCE for land measurement, resource allocation, and astronomy. By the 1st century CE, these traditions had been gathered into *The Nine Chapters on the Mathematical Art* (九章算術, Jiuzhang Suanshu), a handbook of matrix-like techniques for solving linear equations, calculating proportions, and determining areas and volumes.

By the 3rd century CE, the *Chinese remainder theorem* showed that it was possible to determine a unique unknown number from several different clues — each clue being the remainder left when the number was divided by a different base. By carefully scaling and combining these clues, they could be made to match up so that together they revealed the original number. This was a striking demonstration of how separate pieces of partial information could be fitted together to recover a complete answer, and it displayed a deep insight into the structure of numbers.

Liu Hui (circa 225–295 CE) is known through his commentary on *The Nine Chapters on the Mathematical Art*, which he presented to local authorities while serving as a scholar-official responsible for tasks such as land surveying and tax assessment. His work aimed to make existing methods more reliable: he supplied geometric proofs, refined algorithms for solving systems of equations, formalized rules for negative numbers, and introduced clear notation for decimal fractions. His celebrated calculation of π, using a polygon-approximation method akin to Archimedes's exhaustion technique, reflected both ingenuity and a concern for verifiable accuracy in computations that had direct consequences for governance.

Qin Jiushao (1247 – 1324 CE), active during the Song and early Yuan periods, was a high-ranking official whose duties included land surveying, tax assessment, and astronomical forecasting. These responsibilities demanded precise and efficient computation, which shaped his mathematical work. In his *Mathematical Treatise in Nine Sections* (數書九章, Shushu Jiuzhang), Qin developed advanced techniques for handling algebraic equations, breaking them into sequences of simpler, iterative steps to extract roots and coefficients with greater speed and accuracy. His methods, together with earlier Chinese techniques for representing and manipulating unknowns, solving systems of equations, and working with powers, amounted to a fully algebraic tradition — distinct from, and apparently developed independently of, the Mediterranean lineage from Diophantus.

As mathematical traditions evolved independently in the Indian subcontinent, early geometric rules were recorded in *Rules of the Cord* (शुल्बसूत्र, Śulbasūtra) around 500 BCE. Composed for the precise construction of Vedic fire altars, these texts codified proportions, angles, and spatial relations, providing an early framework for geometry in India.

By the 5th century CE, Indian mathematics had undergone a profound transformation with the introduction of zero as a number in its own right, integrated into a decimal place-value system. This advance enabled calculations of unprecedented scope and precision. Aryabhata (circa 476–550 CE) employed this system in his astronomical and trigonometric treatises, applying it to vast numbers and complex computations. Building on this foundation, Brahmagupta (circa 598–668 CE) set out explicit rules for arithmetic involving zero and negative numbers in his *The Correctly Established Doctrine of Brahma* (ब्रह्मस्फुटसिद्धान्त, Brahmasphuṭasiddhānta). He also gave systematic methods for solving quadratic equations.

By the 8th century CE, following the Islamic conquests into the Indian subcontinent, Indian mathematical texts were translated into Arabic. These ideas became part of a wider intellectual exchange that characterized the Islamic Golden Age, roughly during the 8th–14th centuries CE. In Baghdad, the Abbasid Caliphate established the House of Wisdom in the early 9th century, where scholars collected, translated, and expanded upon Greek, Indian, and Babylonian works.

Among its most influential figures was Muhammad ibn Musa al-Khwarizmi (circa 780–850 CE). In *The Compendious Book on Calculation by Completion and Balancing* (كتاب الجبر والمقابلة, Kitāb al-Jabr wa'l-Muqābala), he systematically presented methods for solving linear and quadratic equations, giving the subject its name — *algebra* from 'al-jabr'. His work also contributed to the development of step-by-step computational procedures later known in Europe as *algorithms*, derived from the Latinized form of his name.

Abu Rayhan al-Biruni (circa 973–1048 CE), a prolific mathematician, astronomer, and natural philosopher, refined the use of spherical trigonometry for measuring celestial bodies. He combined geometric reasoning with careful observation to estimate the Earth's radius with remarkable accuracy. His approach exemplified the era's cross-cultural synthesis, drawing on Greek, Persian, and Indian sources and emphasizing empirical verification.

Omar Khayyam (1048 – 1131 CE), remembered in the West as a poet, was also an innovative mathematician and astronomer. He classified cubic equations into distinct types and demonstrated their solutions using conic sections, creating a bridge between algebra and geometry that anticipated Renaissance developments in algebraic theory. He also reformed the Persian calendar, achieving an accuracy comparable to the Gregorian system. His work illustrates the Golden Age ideal: theoretical insight paired with practical application.

The mathematics of the Islamic Golden Age served as a bridge between the ancient world and the emerging intellectual life of medieval Europe. By the 12th century, translation centers in places such as Toledo and Sicily rendered Arabic mathematical and astronomical works — including those of Al-Khwarizmi, Al-Biruni, and Omar Khayyam — into Latin. These texts, themselves syntheses of Greek, Indian, and Babylonian knowledge, introduced European scholars to advanced algebra, refined geometry, and a systematic, empirical approach to investigation that would help shape the scientific thought of the Renaissance.

The translation of both Greek and Arabic works into Latin revitalized European scholarship. Theoretical frameworks from antiquity, combined with the mathematical methods of the Islamic world, created an environment in which logical rigor and quantitative precision were paramount. This was not a simple recovery of lost knowledge but a transformation. Abstract reasoning from the Greeks, algebra and trigonometry from Indian mathematicians, and the synthetic innovations of Islamic scholars merged into a coherent intellectual tradition that defined the Renaissance — an era of scientific inquiry, artistic innovation, and humanistic values grounded in rational thought and empirical evidence.

The practical mathematics of Rome had also shaped this legacy, particularly in Italy. While Roman contributions were less theoretical than Greek ones, they excelled in engineering, surveying, and architecture. In *On Architecture* (De Architectura), Vitruvius (circa 30–15 BCE) codified geometric and proportional principles for buildings, bridges, and aqueducts, ensuring structural stability and visual harmony. Roman engineers developed precise techniques for measurement and construction that Renaissance architects revived and adapted. In the early medieval period, Boethius (circa 480–524 CE) preserved and transmitted Greek mathematical and logical works through translations and commentaries, keeping the tradition alive until its reintroduction during the Renaissance. The combination of Roman practical methods with recovered Greek theory provided a foundation for the intellectual flourishing that followed.

Amid the intellectual revival of the 12th and 13th centuries, Leonardo of Pisa (circa 1170–1250 CE), known as Fibonacci, played a pivotal role in transforming European arithmetic. In *The Book of Calculation* (Liber Abaci, 1202 CE), he introduced the Hindu–Arabic numeral system to a wider European audience, demonstrating through commercial and trade problems how it simplified computation compared to Roman numerals. His text not only popularized a more efficient notation but also showed how abstract numerical relationships could be applied to solve practical problems, laying groundwork for later advances in algebra and number theory.

One of his most famous examples, the "rabbit reproduction problem", asked how many pairs of rabbits would result after one year if each pair reproduced monthly from its second month onward. The rule — that each month's total equals the sum of the previous two months — generated the sequence $1, 1, 2, 3, 5, 8...$, now called the *Fibonacci sequence*. While posed as a theoretical problem, the sequence models idealized exponential growth and appears in diverse natural patterns. When squares with side lengths matching consecutive Fibonacci numbers are placed together and quarter-circles drawn within them, the arcs trace a spiral approximating the golden spiral. Similar forms occur in seashells, hurricanes, and galaxies. In plants, the sequence describes arrangements such as leaves around a stem (*phyllotaxis*) and the seed spirals of a sunflower, patterns that maximize sunlight exposure and packing efficiency.

Elsewhere in Italy, Jordanus de Nemore (circa 1200–1280 CE), or Jordanus Nemorarius, applied mathematical reasoning to problems in mechanics. Though much of his work survives only through later commentary, he is credited with systematic methods for analyzing forces and equilibrium in simple machines, especially levers. By quantifying how force varies with distance — the principle of mechanical advantage — he developed geometric and numerical techniques to locate a lever's precise balance point. These methods, arising from practical needs such as lifting loads and stabilizing structures, formed part of the conceptual foundation for later mechanics and applied mathematics.

At Merton College, Oxford, a group of students later known as the Oxford Calculators — Thomas Bradwardine (circa 1300–1349), William Heytesbury (circa 1313–1372), Richard Swineshead (circa 1320–1360), and John Dumbleton (circa 1310–1349) — was rediscovering Aristotelian thought. Against a backdrop of war, plague, and political instability, they moved from qualitative description to numerical analysis. Their central advance was the treatment of uniformly accelerated motion, which they called *uniform difform motion*, and the formulation of the *mean speed theorem*: a body under constant acceleration travels in a given time the same distance as a body moving uniformly at the average of its initial and final speeds. Using geometric and verbal arguments, they expressed change (motion) with quantitative precision.

Nicole Oresme (circa 1323–1382 CE) was a contemporary French scholar who, in the 14th century, pioneered the use of geometric diagrams to represent the variation of quantities over time — a kind of chart. Motivated by a desire to reconcile and improve upon Aristotelian ideas of motion, Oresme arrived at his own *mean speed theorem*, possibly inspired by the Oxford Calculators. By illustrating how the distance traveled under uniformly accelerated motion could be represented as the area under a visual curve, he anticipated, in a purely geometric way, the concept of integration that would later be formalized in calculus. His work, which explored the relationships between time, speed, and distance, was driven by both theoretical curiosity and practical needs, such as understanding natural phenomena and improving mechanical devices. Oresme's innovative approach transformed qualitative observations into quantitative analyses, thereby bridging medieval empiricism and the analytical methods of the Renaissance.

The Italian Filippo Brunelleschi (circa 1377–1446) played a central role in formulating linear perspective, reshaping the depiction of space in art and architecture. In Florence, he conducted demonstrations — including the use of mirrors to compare a painted panel with the real view — to show that a flat surface could reproduce depth by rendering parallel lines as converging at a single point. Building on these empirical insights, Leon Battista Alberti (circa 1404–1472) authored the influential treatise *On Painting* (De pictura, circa 1435), in which he systematically explained the mathematical principles behind perspective. Alberti introduced concepts such as the *horizon line* and *vanishing point*, and he provided step-by-step guidelines for proportionally scaling objects in a painting to achieve a realistic depiction of three-dimensional space on a two-dimensional surface.

Alberti is often regarded as a "father of Western cryptography" for his invention of the cipher disk, in order to secure communication in the politically turbulent environment of Renaissance Italy. The cipher disk consisted of two concentric disks: the outer disk was typically fixed with the standard alphabet in order, while the inner disk could be rotated to align a scrambled or keyed version of the alphabet with the outer one. This configuration allowed the user to substitute letters according to a shifting pattern; by rotating the inner disk, the cipher would change continuously throughout the message. In effect, this meant that the same letter in the plaintext could be encoded differently depending on its position in the message. The ability to rapidly change cipher alphabets using his disk provided a practical and effective method to protect sensitive diplomatic or military information.

Johannes Müller von Königsberg (circa 1436–1476), also known as Regiomontanus, was a mathematician and astronomer of early Renaissance Germany. He refined methods for relating the sides and angles of triangles, recognized their connection to circle geometry, and compiled tables that allowed these ratios to be calculated for any given triangle. Such tools were essential for predicting the positions of celestial bodies and for accurate surveying, and became a foundation for trigonometry in navigation, astronomy, and engineering.

One Polish mathematician and astronomer who would later own and annotate a copy of Regiomontanus's work was Nicolaus Copernicus (circa 1473–1543). In *On the Revolutions of the Heavenly Spheres* (De revolutionibus orbium coelestium, 1543), Copernicus placed the Sun, rather than the Earth, at the center of the universe. Although he retained circular orbits, his model offered a more coherent and mathematically unified explanation of planetary motions than Ptolemy's geocentric system of epicycles and deferents. In an early manuscript, Copernicus credited Aristarchus of Samos with first proposing a moving Earth, but this acknowledgement was removed before publication, leaving only indirect traces of Aristarchus's influence. The reason for its removal is unknown, but the connection between the ancient proposal and Copernicus's reformulation remains evident.

Francesco Maurolico (1494–1575), a Sicilian mathematician deeply engaged in the Renaissance recovery of classical Greek learning, made a lasting contribution to the methodology of proof with the first explicit and systematic formulation of mathematical *induction*. While his reconstructions and extensions of Archimedes's works were substantial, it was his presentation of induction in *Two Books of Arithmetic* (Arithmeticorum libri duo, 1575) that marked a decisive advance. Earlier mathematicians had reasoned inductively, but Maurolico formalized the process: first establish a base case, then prove that if a proposition holds for an arbitrary integer k it must also hold for $k + 1$. Applied step by step, this shows the statement must be true for all natural numbers. He used the method to show, for example, that the sum of the first n odd integers equals n^2. This allowed mathematicians to go from checking a few specific cases to proving a statement for every number without exception — part of a wider Renaissance movement toward abstract methods capable of handling whole classes of problems rather than isolated examples.

In antiquity, problems were typically expressed in words or through geometric constructions aimed at particular examples — calculating a specific area or volume, for instance — without a notation that could address the general case. In the Renaissance, alongside developments like Maurolico's induction, mathematicians increasingly sought techniques and symbolic systems that could describe and solve entire families of equations. This evolution of algebra from rhetorical and geometric forms into a symbolic language marked a decisive shift from ancient approaches, paving the way for methods that could unify disparate problems under a common framework.

In the years that followed, changes in mathematical notation and calculation made problems easier to write down, understand, and solve. Robert Recorde (circa 1510–1558), in *The Whetstone of Witte* (1557), introduced the equals sign (=) to the English-speaking world, explaining his choice of two parallel lines because "noe 2 thynges can be moare equalle". Before this, equality was usually written out in words, slowing down the flow of algebraic work. Simon Stevin (1548–1620) promoted decimal fractions in *The Tenth* ("De Thiende", 1585), showing how any quantity — from money to measurements — could be expressed in base ten notation and handled with the same rules as whole numbers. This greatly simplified calculations in trade, surveying, and astronomy.

William Oughtred (1574–1660) further advanced mathematical notation in *The Key to Mathematics* (Clavis Mathematicae, 1631), popularizing × for multiplication and :: for proportion. Simultaneously, *logarithms* emerged to relieve the burden of long multiplications in astronomy and navigation. John Napier (1550–1617) constructed tables pairing each number with a companion value — its logarithm — so that a product could be found by adding the logarithms and then reversing the lookup in the same tables. Jost Bürgi (1552–1632) developed related tables independently, and Henry Briggs (1561–1630) worked with Napier to recalibrate them to base ten "common" logarithms that fit Stevin's decimals. Oughtred's *slide rule* embodied this process physically: placing two logarithmic scales against each other so that sliding to add distances corresponds to adding logarithms, then reading back gives the desired number. The tool's clarity depended on exponents. Michael Stifel (circa 1487–1567) treated powers more systematically, and René Descartes (1596–1650) introduced modern exponent notation such as x^3. Together, the equals sign, decimal fractions, logarithms, the slide rule, and exponent notation revolutionized the language of mathematics and calculation.

Beyond individual operators, symbolic manipulation — the representation of numbers, unknowns, and operations with symbols — has proven time and again to be a powerful tool for advancing mathematics. By replacing verbose, case-by-case descriptions with concise notation, mathematicians could express general rules and patterns that applied to entire classes of problems rather than single instances. Using a letter to represent an unknown made it possible to write down an equation and solve it by applying well-defined rules, revealing structures that would be obscured in verbal form. This abstraction allowed ideas to be combined in new ways and extended beyond their original contexts.

The transition from rhetorical or geometric methods to symbolic algebra during the Renaissance marked a decisive shift. Scipione del Ferro (1465–1526), Niccolò Tartaglia (1499–1557), and Girolamo Cardano (1501–1576) solved third and fourth degree equations using elaborate substitutions and transformations, recording their methods in full sentences — for example: "Take the unknown, multiply it by a certain number, and add another number, so that the whole sum is nothing". François Viète (1540–1603) introduced systematic notation for quantities and operations, making such results easier to state, compare, and generalize. This evolution of symbolic manipulation laid the foundation for modern algebra by providing a language capable of capturing the essential features of mathematical problems in a precise, reproducible way.

In the early 16th century, del Ferro discovered a method for solving a special form of the cubic (third degree) equation in which the quadratic (second degree) term is missing, but he kept it secret. His method became partly known through Tartaglia, who, in public mathematical contests, demonstrated a way to transform a cubic into a simpler problem — finding two numbers whose cubes sum to a constant — reducing it to a form governed by quadratic relationships. Cardano learned of Tartaglia's method and published a general solution for cubic equations in *The Great Art* (Ars Magna, 1545), extending it to *quartic* (fourth degree) equations and introducing the notion of *imaginary* numbers when square roots of negative numbers appeared in intermediate steps. Cardano noted that such quantities could cancel out to yield real answers, a paradox that would later inspire deeper study.

Cardano's student Lodovico Ferrari (circa 1522–1565) devised the first general solution for quartic equations. His approach reduced the quartic to a cubic, then factored it into two quadratics solvable by the familiar quadratic formula. With this, mathematicians of the Renaissance could solve all *polynomial equations* up to degree four using systematic methods. Despite hopes that similar substitutions might unlock the *quintic* (fifth degree) and higher degrees, no general formula was found. The methods worked for lower degrees because the relationships among their solutions depend on only a few independent parameters; by the fifth degree, the complexity of these relationships prevents expressing the solution as a finite sequence of basic arithmetic operations. Much later, in the 19th century, mathematicians developed a deeper theory — *group theory* — which showed that for quintic equations the process for finding solutions cannot be broken down into simple operations the way it is for lower-degree equations.

The nature of this barrier can be understood by analogy with geometry. For equations of degree one through four, the relationships among their solutions can be represented within a framework that corresponds to spatial structures familiar from one, two, or three dimensions. The structure is simple enough that it can be unravelled using a finite sequence of basic operations, just as a three-dimensional object can be fully described by its two- and one-dimensional features. A quintic equation introduces an additional degree of freedom: the interdependencies among its five solutions cannot be compressed into the simpler, lower-dimensional framework. This is akin to the difficulty of picturing a four-dimensional object — spatial intuition shaped by three dimensions cannot capture the full complexity. Just as visualizing higher dimensions requires different mathematical models, solving the general quintic required entirely different approaches from abstract algebra.

The puzzling appearance of square roots of negative numbers in Cardano's formulas prompted Rafael Bombelli (1526–1572) to investigate them systematically. In *Algebra* (1572), he established consistent rules for adding, multiplying, and extracting roots of these so-called imaginary numbers, showing they could be handled reliably and used to solve otherwise intractable problems. Though met with initial skepticism, Bombelli's work laid the foundation for the modern theory of complex numbers.

Ostilio Ricci (circa 1540–1603) further developed and taught algebraic methods at universities such as Pisa and Padua, embedding symbolic algebra into mathematical education. This abstract approach to manipulating equations — shifting from case-specific procedures to a general symbolic system — wove itself into the intellectual fabric of the time and directly influenced young scholars such as Galileo Galilei (1564–1642).

Galileo, with his training in symbolic algebra, revolutionized the understanding of nature by not only asserting that natural phenomena obey mathematical laws, but also demonstrating this through carefully designed experiments and analytical methods. Using inclined planes to slow the motion of falling bodies, he was able to measure both time and distance accurately. Rolling balls down a smooth slope and timing them with water clocks or pendulum-based devices, he found that the distance travelled was proportional to the square of the elapsed time — evidence of uniform acceleration. If distance had been proportional to time, velocity would have been constant; the square relationship showed that velocity increased steadily. Marked intervals along the plane provided spatial measurements, and repeated trials reduced error, yielding consistent results.

Galileo applied the same combination of careful measurement and mathematical analysis to astronomy. After improving the telescope to achieve twenty-times magnification, he began, in 1609, a program of systematic observation. He recorded the Moon's rugged surface, the phases of Venus, and four satellites orbiting Jupiter — discoveries that challenged the Aristotelian and geocentric worldview. By tracking the motions of these bodies and applying geometric reasoning, he demonstrated that celestial objects obeyed the same mathematical principles as terrestrial ones.

Nightly observations of Jupiter revealed four small moons circling the planet in regular orbits. Venus displayed a cycle of phases, from crescent to full, that matched its changing position relative to the Sun — possible only if Venus orbited the Sun. These findings provided direct, quantitative evidence for a heliocentric system.

Building on the Copernican model, Johannes Kepler (1571–1630) transformed it into a more accurate description of planetary motion using positional data meticulously collected by Tycho Brahe (1546–1601). Tycho's observatories, including Uraniborg on the island of Hven, produced measurements of stars, planets, and comets precise enough to expose subtle deviations in planetary paths. Kepler's study of Mars showed that no circular model, however modified, could fit the observations. He found instead that an ellipse — a stretched circle defined by two foci — matched the data exactly. This became his *First Law of Planetary Motion*: planets move in elliptical orbits with the Sun at one focus. His *Second Law of Planetary Motion*, the *Law of Equal Areas*, states that a line joining a planet to the Sun sweeps out equal areas in equal times, requiring faster motion when closer to the Sun and slower motion when farther away. His *Third Law of Planetary Motion*, the *Harmonic Law*, establishes that the square of a planet's orbital period is proportional to the cube of its average distance from the Sun — the same ratio for all planets. Though still couched in geometric language and ratios, Kepler's work replaced the circular and epicyclic schemes of antiquity with a precise, predictive mathematical model.

In the generation following Galileo's experiments and Kepler's planetary laws, Marin Mersenne (1588–1648) became a central link among Europe's leading scientific minds — Galileo himself, René Descartes, Pierre de Fermat, Blaise Pascal, Christiaan Huygens, Evangelista Torricelli, Thomas Hobbes, and many others. A Minim friar in the Order of Minims founded by Saint Francis of Paola, Mersenne lived under a rule that combined humility and contemplation with encouragement of scholarly work, particularly in mathematics and science. Fascinated by perfect numbers and primes of the form $2^n - 1$, he pursued these problems with dedication. More influential still, from about 1620 until his death he served as Europe's scientific 'postbox', translating and circulating results, relaying problems between correspondents, and distilling complex findings — including Kepler's astronomical work — at a time before formal scientific journals.

René Descartes (1596–1650), in *Geometry* (La Géométrie, 1637), introduced a coordinate method in which every point in a plane could be described by a pair of numbers. This was not merely a rephrasing of known ideas — it was a fundamental shift in perspective. By fixing an origin and perpendicular axes, curves could be expressed as equations in variables such as x and y. Lines, circles, and more complex curves could be described by algebraic relationships, and problems such as finding the intersection of a circle and a line could be reduced to solving an equation. Using familiar operations — addition, subtraction, multiplication, division, and extraction of roots — mathematicians could determine exact coordinates for points that geometry alone would have located only by construction. This unification of geometry and algebra created a symbolic language capable of representing both figures and formulae, forming a foundation for later analysis and for the mathematical treatment of physical problems.

The success of this algebraic framework accelerated a wider turn toward abstraction in European mathematics: symbolic manipulation of equations, investigation of infinitesimals, and summation of infinite series. These methods vastly extended the scope of mathematics but also raised the threshold of technical entry, widening the gap between formal theory and direct physical observation.

It was in that observational realm — tangible phenomena measured with care — that the next decisive advances would arise. In subjects like electricity and magnetism, little touched by the prevailing analytic tools, progress would depend on experiment and visualization — above all in the work of Michael Faraday, who, despite a perceived lack of mathematical sophistication, possessed an unparalleled intuitive genius and experimental prowess.

Intuition Leads the Way

In 1813, at the age of twenty-one, Michael Faraday began his intensive study of chemistry as an assistant to Humphry Davy (1778–1829) at the Royal Institution of Great Britain. It was his first sustained experience in a laboratory, and at the outset his duties were basic — washing bottles, mixing and storing chemicals, and keeping the glassware and instruments in good order for Davy's experiments.

Faraday was born in Newington, on the southern outskirts of London, into a working family of limited means. His father, James Faraday, was a blacksmith whose chronic ill health often kept the household in financial difficulty. With no state provision for schooling and no money for private education, Michael received only a few years of basic instruction in reading, writing, and arithmetic before needing to earn his keep. He first worked as a delivery boy for a local bookseller and stationer in London, who also ran a bookbinding business. In 1805, at the age of fourteen, Faraday was apprenticed for a seven-year term — a common path for working-class boys, offering training, board, and a modest allowance.

The trade gave Faraday an unexpected education. He was allowed to read the books he bound and sold, and Faraday eagerly absorbed works such as *Conversations on Chemistry* (1805) by Jane Marcet (1769–1858), which he later recalled was especially influential. Faraday's voracious reading and determination allowed him to absorb concepts far beyond his formal schooling. By his late teens, his interest in science led him to attend public lectures at the Royal Institution, where live demonstrations of electrical and chemical phenomena left a lasting impression. In 1812, he took detailed notes at a series of lectures by the renowned chemist Humphry Davy, presenting them to the chemist along with a request for employment. Impressed by Faraday's keen observations and self-taught knowledge, Davy offered him a position the following year. The age of science brought fresh opportunity in every direction, as it had for Davy himself.

In the years before Humphry Davy emerged, the field of chemistry was undergoing a profound transformation, largely driven by the work of Antoine-Laurent Lavoisier (1743–1794), a central figure in what became known as the Chemical Revolution. Lavoisier, in his landmark textbook *Elements of Chemistry* (Traité Élémentaire de Chimie, 1789), redefined the foundations of the field. He shifted the focus from qualitative observations to precise, quantitative measurements, emphasizing the importance of mass conservation in chemical reactions. Robert Boyle (1627–1691) had earlier argued in *The Sceptical Chymist* (1661) that an element should be understood as a substance not yet resolved into simpler bodies. Lavoisier adopted this pragmatic definition and published a list of such substances. Together with Louis-Bernard Guyton de Morveau (1737–1816), Claude-Louis Berthollet (1748–1822), and Antoine-François Fourcroy (1755–1809), he introduced a systematic nomenclature in *Method of Chemical Nomenclature* (Méthode de nomenclature chimique, 1787), displacing much alchemical vocabulary. Not just a scientist; Lavoisier was also a powerful member of the Ferme Générale, an outsourced private tax company notorious for its exploitative practices under the French monarchy. Accused of financial crimes against the people, Lavoisier was executed by guillotine in 1794 during the French Revolution.

Following Lavoisier's transformation of chemistry into a quantitative science, Charles-Augustin de Coulomb (1736–1806) emerged as a key figure in the study of electricity. Although not directly connected to Lavoisier, Coulomb's education at a French military engineering academy instilled in him the same emphasis on precision measurement that defined the scientific method. His most important work on electrostatics relied on the torsion balance: a lightweight horizontal rod suspended by a thin, calibrated fiber that resisted twisting. At one end of the rod he mounted a small metal sphere, charged by rubbing it with materials such as wool or silk, a common technique of the period. Introducing a second charged sphere at varying distances caused the rod to twist; by carefully measuring the angle of torsion as he altered the separation, Coulomb showed that the force was directly proportional to the product of the charges and inversely proportional to the square of the distance between them. This clear quantitative law paralleled the kind of numerical relationships Lavoisier had sought in chemistry.

While Coulomb was refining electrical measurement in France, developments in Chemistry were propelling Humphry Davy (1778–1829) to prominence in Britain. Born in Penzance, Cornwall, Davy was apprenticed to a surgeon–apothecary, where he began to teach himself chemistry. He studied Lavoisier's publications intensely and soon found an opportunity at the Pneumatic Institution in Bristol, where he investigated the physiological effects of gases. His daring experiments with nitrous oxide — including self-administration and public demonstrations of its intoxicating "laughing gas" effects — gained him notoriety. In 1799 he published *Researches, Chemical and Philosophical; Chiefly Concerning Nitrous Oxide*, a work combining detailed experimentation with youthful enthusiasm. Its reception brought him to London in 1801 as a lecturer at the newly founded Royal Institution, a platform that established his reputation and allowed him to pursue groundbreaking electrochemical research.

Davy's work with the *galvanic battery* (also known as the *voltaic pile*) was central to this research. Scaling up Alessandro Volta's original design, he built batteries with hundreds, and eventually thousands, of zinc–copper pairs to deliver high currents. With these, he explored *electrolysis*: the chemical decomposition driven by electric current. Recognizing that aqueous solutions often decomposed water before the target compound, Davy used molten salts such as potassium hydroxide or sodium hydroxide, heated until liquid. Employing inert platinum electrodes, he passed current directly through the molten substance, producing silvery, highly reactive metallic globules — potassium or sodium, depending on the electrolyte. By collecting these metals under oil to prevent immediate reaction with air or water, and observing their reversible formation from hydroxides, Davy confirmed both their elemental nature and the general principles of electrolysis: that electricity could act as a precise agent of chemical change.

His public lectures at the Royal Institution became famous for combining scientific clarity with dramatic effect. In a tiered auditorium, the central table would be set with polished Leyden jars, intricately arranged wires, and freshly assembled voltaic piles of alternating zinc and copper plates. Davy explained each component in clear terms before building suspense with a carefully timed demonstration. Charging a Leyden jar from the battery, he would produce a sudden, brilliant spark — a miniature bolt of lightning — accompanied by a sharp crack that echoed through the room. In another lecture, he decomposed water in a clear vessel by electrolysis, streams of hydrogen and oxygen bubbles forming silently at the electrodes. Contemporary accounts noted that Davy's descriptions were as vivid as his demonstrations, likening the rising bubbles to the interplay of unseen natural forces.

For audiences, these events were both instruction and spectacle, and among the most attentive was the young Michael Faraday. His access to such lectures, often free to the public, was the product of determination and the Royal Institution's openness to all social classes. Faraday took meticulous notes, especially during a series on high-current electrolysis, recording apparatus configurations, procedures, and results in detail. A later report in the Journal of the Royal Institution noted that Davy privately remarked on the precision of Faraday's observations and the acuity of his questions. This recognition was what led Davy to invite Faraday into his laboratory in 1813.

On joining the laboratory, Faraday was immersed in a regime where even routine work demanded precision. His duties ranged from cleaning and assembling apparatus to preparing chemicals and logging observations in his ever-present notebooks — a habit he credited with sharpening his experimental focus. Under Davy's close supervision, he learned the practical challenges of maintaining series-connected voltaic cells, adjusting currents, and handling reactive materials. Within months he was entrusted with more complex tasks, such as refining cell configurations and improving electrolysis procedures. By the mid-1810s, according to contemporaries and his own memoirs, Faraday was designing and executing parts of Davy's experiments, optimizing apparatus, and assisting in public lectures — sometimes explaining procedures himself. By 1820 he had moved from assistant to active collaborator, contributing original insights that shaped the laboratory's progress.

In the spring of 1820, Hans Christian Ørsted (1777–1851) announced that an electric current could deflect a magnetic needle, firmly establishing a direct link between electricity and magnetism. The news sparked an immediate surge of investigation across Europe as scientists sought to quantify and explain this magnetic effect of electricity.

Among the earliest and most prominent were André-Marie Ampère (1775–1836) and the team of Jean-Baptiste Biot (1774–1862) with Félix Savart (1791–1841). Ampère concentrated on the overall interactions between current-carrying conductors. Through systematic experiments, he showed that two parallel wires carrying currents in the same direction attract, while those with currents in opposite directions repel. The strength of this force depended on the magnitude of the currents and their relative orientation. From these observations, Ampère formulated consistent rules describing the magnetic interaction between conductors, establishing a clear quantitative relationship between current and magnetic force.

Biot and Savart approached the problem at a smaller scale. They examined the local magnetic influence of a short segment of current-carrying wire, measuring how the field's strength varied with current, distance, and orientation. Their experiments led to a mathematical formulation that described the spatial variation of the magnetic field around a conductor, complementing Ampère's broader results with a precise description of the field due to an element of current.

Ørsted's discovery reached England more slowly than the Continent. In November 1820, entries in Michael Faraday's laboratory notebooks — preserved in the Royal Institution archives — show that within weeks of hearing the news, he constructed a simple apparatus to replicate the effect. Placing a current-carrying wire near a magnetic compass, he observed the needle's deflection.

Over the following months, and into early 1821, Faraday's work shifted from replication to systematic investigation. He altered the distance between wire and compass, varied their orientation, and experimented with different voltaic pile configurations to adjust current strength. These methodical changes allowed him to begin quantifying the link between current and magnetic deflection, planting the first seeds of a line of inquiry that would occupy him for the next decade. Although still working under Davy's general supervision, these experiments were increasingly conceived and directed by Faraday himself. Surviving correspondence from within Davy's circle suggests that Davy was impressed by Faraday's precision and recognized the potential for a deeper unification of electrical and magnetic phenomena.

By 1822, Faraday was setting his own research agenda at the Royal Institution. Davy remained an influential figure, but the younger scientist was now pursuing independent lines of inquiry. That year, he noted in his diary his goal: "Convert magnetism into electricity". He constructed electromagnets and solenoids of varied coil arrangements and core materials, exploring how the strength and geometry of magnetic fields could be controlled. His persistence in studying the interaction between magnets and conductors gradually revealed the principle that a changing magnetic field can induce an electric current in a nearby conductor.

This work culminated in 1831, fulfilling the aim he had set nearly a decade earlier. In August, Faraday wound two coils of wire around an iron ring, connecting one coil to a battery. When the battery circuit was made or broken, a transient current appeared in the second coil, showing that only a changing current — and thus a changing magnetic field — could produce induction. In October, he demonstrated that moving a bar magnet through a coil could generate a continuous current, as detected by a *galvanometer* — an instrument that detected small currents via the deflection of a needle in a magnetic field. Soon after this experiment, Faraday devised an electric generator, known as the "Faraday disc": a copper disc rotating between the poles of a horseshoe magnet to produce a steady current. In the same productive period, he also carried out experiments that led to the liquefaction of chlorine and the discovery of benzene — two unrelated but nonetheless significant achievements.

While Faraday was refining his understanding of induction in England, Joseph Henry (1797–1878) in the United States was advancing parallel lines of inquiry. In the early 1830s, Henry conducted experiments that confirmed the principles of electromagnetic induction and revealed the phenomenon of self-induction — the influence of a circuit's own magnetic field on its current. His work, involving the construction of powerful electromagnets and the careful measurement of induced effects, complemented Faraday's conceptually driven studies by emphasizing quantitative and practical aspects of electromagnetic principles.

At nearly the same time, theoretical refinements were shaping the field. In 1834, the Estonian physicist Emil Lenz (1804–1865) articulated what became known as *Lenz's law*: the induced current in a circuit always flows in a direction that opposes the change in magnetic flux that produced it, an expression of energy conservation in electromagnetic systems.

Meanwhile, in Göttingen, Carl Friedrich Gauss (1777–1855) was thinking about how to express, in the simplest possible terms, the connection between a source — such as an electric charge — and the way its influence fills the space around it. He imagined enclosing the source inside an invisible bubble, a perfectly smooth closed surface that could be of any size or shape, so long as it surrounded the charge completely. From the source, the *field* — the force felt at each point in space — radiates evenly in every direction, like light from a single point. The farther the surface is from the source, the more thinly that influence is spread; its strength diminishes with the square of the distance, as Coulomb had shown.

At the same time, as the enclosing surface is pushed farther out, its area grows, increasing in proportion to the square of the distance. One effect weakens the field, the other enlarges the surface to be filled. Gauss saw that these two changes balance each other perfectly: when the surface's radius is doubled, the field at each point is four times weaker, but the surface area is four times greater. The total 'amount' of field crossing the surface remains unchanged.

This total is what came to be called the *flux* — a term borrowed from the language of flowing fluids, from the Latin *fluxus*, meaning 'flow'. Regardless of the size or shape of the surface, so long as it encloses the source entirely, the total flux is the same, determined solely by the strength of the source within. Gauss gave this relationship a precise mathematical form, creating a most compact and far-reaching statement about how fields behave in space.

Faraday's path into this same conceptual territory was by experiment rather than abstract reasoning. In his 1831 paper *Experimental Researches in Electricity*, he insisted that "the lines of magnetic force" were not mental constructs but real, physical entities: "the lines of force are not lines drawn by the mind but a description of the continuous action emanating from the magnet". By 1832 and in later writings, he stressed that while mathematics could describe field behavior, it could not fully convey the lived reality of such phenomena: "Mathematics may outline the laws governing these forces, yet it is through direct observation of their effects in nature that one truly apprehends their continuous and dynamic interplay".

Since Faraday regarded the "lines of magnetic force" as real structures in space, he reasoned, it should be the same whether produced by a frictional machine, a chemical battery, or electromagnetic induction. To test this unity, he needed a process that could be driven by each form of electricity and measured with precision. Electrolysis offered exactly that. Passing a current through a solution produced tangible, weighable chemical changes, providing a common measure for all sources. From 1832 to 1834, Faraday would conduct experiments to test his hypothesis.

For these experiments, Faraday generated and measured electrical charge, devising methods to quantify the resulting chemical changes with great care. He weighed electrodes before and after electrolysis to detect minute changes in mass from deposited metals. When gases were produced, he measured their volumes with a *eudiometer* — a graduated glass tube in which the gas displaced water — a device used earlier by chemists such as Joseph Priestley (1733–1804).

Faraday compared the continuous current from a voltaic pile with the brief discharge from a Leyden jar, monitoring both with a galvanometer. He charged the jars using static electricity from friction, with the jar's outer coating connected to ground during charging to complete the circuit and allow full potential to build on the inner coating. Once fully charged — indicated when further charging no longer increased the galvanometer deflection during test discharges — the jar was discharged through the electrolysis apparatus. A single discharge caused only a tiny chemical change, so Faraday inserted high resistance into the circuit to slow the current for measurement, and repeated the process many times to accumulate a measurable effect. He estimated the total charge by summing the current pulses over their durations and compared the resulting chemical changes with those from continuous current. From this, he showed that the same proportional relationship between electricity and chemical change held for all sources.

Several years earlier, in 1827, Georg Simon Ohm (1789–1854) had set out a mathematical law describing how an electrical current depends on the driving potential and the resistance of its path. Faraday did not work from Ohm's formulation, but in retrospect their findings complemented each other: Ohm provided a general balance equation for current in conductors, and Faraday demonstrated that, whatever the source, a given quantity of charge produced the same effect.

Between 1834 and 1838, Faraday turned his attention to investigating how insulating materials affect electrostatic induction. At the time, *insulators* — then called *non-conductors* — were generally thought to be inert, merely preventing charges from leaking away. Guided by his belief in the reality of *lines of force*, he reasoned that these lines must traverse the space between charged bodies. If that space contained matter, the lines would have to interact with it somehow.

To explore this, he needed a field source that would remain unchanged while he varied the material in its path. He chose a charged glass rod. As an insulator itself, its surface charge was fixed in place and would not redistribute in response to the sample being tested. On the far side of the test position he suspended a polished metal sphere from an insulating thread, with a "pith-ball electroscope" in contact to register changes in the sphere's electrical state. In response to any changes, the *pith ball* — a tiny, lightweight, insulating ball made from the soft, spongy tissue of certain plants — would pick up some of the sphere's displaced charge and swing outward under electrostatic repulsion, giving a clear visible sign of the change.

With only air between rod and sphere, bringing the charged rod close caused the pith ball to deflect. When Faraday inserted plates of different insulating substances into the gap, the amount of deflection changed. This showed that the medium between source and conductor was altering the induction — evidence that the lines of force were interacting with the material through which they passed.

Having demonstrated the effect qualitatively, Faraday turned to quantitative measurement with a Leyden jar. The jar's two metal coatings, separated by an insulating layer, could store a measurable charge at a given potential. By replacing the insulating layer with different materials, he found that the jar's capacity changed systematically. His interpretation was that the dielectric material itself became slightly "charged within" — its particles displaced or oriented so as to set up an internal electrical state opposed to the applied force.

In 1837, Faraday began using the term *dielectric* for such materials. The prefix 'dia-' ('through' and 'opposite') reflected his view that the medium transmits the electrical influence while mounting an internal opposition to it. He called this response *polarization*. In Faraday's mind, this was simply the natural consequence of treating lines of force as real: if they pass through matter, the matter must answer back.

This concept of polarization, though born from electrical experiments, fit neatly into Faraday's broader philosophy that the various natural forces — electricity, magnetism, and perhaps even light — were interconnected manifestations of a common principle. By the mid-1830s, his notebooks and correspondence show him speculating about possible links between magnetism and the behavior of light.

The word *polarization* already carried a suggestive history. In 1808, the French engineer Étienne-Louis Malus (1775-1812) examined sunlight with *Iceland spar*, a crystal long known to split ordinary sunlight into two beams of equal brightness. With ordinary sunlight, rotating the crystal did not change that balance: at every angle the two beams remained equally bright. Malus then looked at indirect sunlight, being reflected off glass at a particular angle. Here rotation of the crystal did matter: as he turned it, the relative brightness of the two beams alternated in a regular pattern every quarter-turn. From these observations, Malus reasoned that the difference lay in the light, not the crystal. If the crystal alone were responsible, rotating it would have altered the beams for ordinary sunlight as well. Because only the reflected beam showed the alternating pattern, he concluded that reflection had imprinted on it an intrinsic directional property that ordinary sunlight did not possess. He called this property *polarization*, borrowing the term from magnetism with its two opposed poles. The name did not assert that light had literal magnetic poles; it marked a directional character revealed by experiment.

For a man of Faraday's sensitivity to language, Malus's choice of term likely seemed more than a metaphor. It hinted at a deeper physical kinship. If light could be polarized in a way that echoed magnetic polarity, perhaps the analogy literally reflected a real connection between light and electromagnetism. One way or another, Faraday was convinced a unifying connection must permeate the natural world.

Guided by this thought, Faraday spent years trying to detect such a link. In the early 1840s, he began by testing whether electric fields could affect polarized light. He passed beams of polarized light through electrolytic solutions under strong electric fields, varying the voltage, the path length, and the solution, but found no measurable change. His notebooks from this period record meticulous adjustments, mounting frustration, and finally the decision to abandon the electric route.

He then investigated the effect of magnetism, setting up experiments with polarized light passing through transparent materials placed between the poles of electromagnets. Again he varied every parameter he could think of — field strength, orientation, material, and the alignment of polarizing prisms — but glass after glass yielded nothing. For years, the effort seemed as fruitless as the electrical tests.

The breakthrough came in 1845, using a dense, optically clear "heavy" glass he had made in the late 1820s while experimenting with glass compositions for telescopes and scientific instruments. This glass, composed of silica, boracic acid, and lead oxide, was exceptionally transparent and optically clear. Faraday polarized a beam of sunlight with a *Nicol prism* — a calcite device introduced by William Nicol (1768–1851) in 1828 that transmits light in one polarization and suppresses the opposite kind. He directed this beam through his heavy glass, positioned between the poles of a powerful electromagnet, and examined it with a second Nicol prism, set to block the polarized beam completely. With the magnet turned off, the beam vanished in total darkness. When he switched the magnet on, the beam reappeared faintly; rotating the analyzer prism could once again extinguish it. The magnetic field had caused a small rotation in the "plane of polarization" — just enough to turn darkness into light.

To Faraday, this was decisive: magnetism could influence light. In his notes he declared he had "magnetized a ray of light". Strictly speaking, the experiment showed that the magnetic field altered the optical properties of the glass, which in turn changed the light's polarization; whether the effect was a direct action on light itself or mediated entirely by the material was not resolved. But for Faraday, convinced of the unity of natural forces, the result was enough to claim a fundamental connection.

Encouraged by his discovery of the *Faraday effect*, Faraday extended his investigations to ask whether all materials might be influenced by magnetism. The possibility had precedent. François Arago (1796–1853), while studying the behavior of magnetic needles under various conditions, found that a rotating copper disk beneath a suspended magnetic needle would cause the needle to turn with the disk, even though copper shows no obvious attraction to a magnet. He carried out similar tests with other metals such as brass and zinc, and with non-metals including glass and wood, noting that a moving body of these substances could still affect a magnetic needle's position. Arago also reported that when certain substances were brought close to a magnetized needle, the needle's motion or final position could be altered, the strength and character of the effect depending on the material. Although he did not attempt to organize these observations into a general classification, they showed that interaction with magnetic fields was not limited to a small group of metals, which he presented in a memoir to the French Academy of Sciences in 1824.

The results were striking. Every material he tested exhibited some form of magnetic response. Iron and similar metals showed the familiar strong attraction, which he termed *paramagnetism*. Most substances, however — including glass, water, wood, and even gases — responded in the opposite way, being repelled by strong magnetic fields. These subtle interactions required very strong magnets — achievable only because Faraday had already developed powerful electromagnets strong enough to reveal these otherwise imperceptible effects. For this universal property, Faraday introduced the term *diamagnetism*.

As with his earlier *dielectric*, these names were chosen with precision. The established term *magnetism* he reserved for the persistent, polarized form of the phenomenon: objects with fixed north and south poles, whose internal magnetic orientation was tied to their geometry and could only be altered by physically reorienting the object or otherwise disturbing its structure.

Diamagnetism, by contrast, was transient and opposed to the applied influence. Borrowing again from the Greek 'dia-' ('through' and 'opposite'), Faraday used the term to emphasize that such materials do not simply allow magnetic fields to pass through them, but develop an internal state diametrically opposed to the external field, producing repulsion in all orientations. The complementary term paramagnetism, from 'para-' ('beside' or 'alongside'), he applied to materials whose internal magnetic "moments" align with an external field, producing attraction. Like diamagnets, paramagnets lacked any inherent or fixed polarity, but unlike diamagnets they reinforced rather than resisted the applied field.

This division of magnetic behavior is perhaps best understood in terms of the internal structure each form requires. Diamagnetism, present in all substances, is electromagnetic induction at the smallest scale: countless fleeting loops of current form in opposition to a change in the magnetic environment, and vanish as soon as that change reverts. Because this response needs no special organization beyond the basic constitution of matter, it is universal. Paramagnetism demands more: tiny magnets already present in the material are free to temporarily turn and align with an applied magnetic field, and return to disorder when that field is removed. Such freedom is absent in most materials, making this behavior less common. Permanent magnetism requires the most exacting arrangement: tiny magnets must not only exist and be able to align, but remain locked in place by the geometry of the material so their orientation persists indefinitely. The relative frequency of these behaviors in nature follows directly from these structural demands — diamagnetism in all matter, paramagnetism in a smaller fraction, and permanent magnetism in only a select few.

In the years following his discoveries of diamagnetism and the Faraday effect, Faraday continued to deepen and clarify the conceptual foundations of electromagnetism. Though declining health limited the pace of his laboratory work in later years, he remained actively engaged in developing a cohesive framework built around his intuitive model of lines of force. He refined this into the idea of a universal, continuous medium filling all space, linking electrical, magnetic, and optical phenomena. These philosophical and conceptual insights formed the direct intellectual groundwork for the electromagnetic theory James Clerk Maxwell (1831–1879) would soon express in mathematical form.

Faraday's lines of force were described vividly and concretely, yet they remained largely qualitative — conveyed through analogy and imagery rather than precise equations. Translating them into a predictive science required a language capable of representing continuous distributions of force throughout space. That language was mathematics, and central to it was the concept of the infinitesimal: analyzing phenomena as the sum of infinitely small elements of space and time, then integrating them into a complete description of the whole. This is exactly what the burgeoning field of *calculus* was designed to do.

But adopting calculus was more than a technical choice. It marked a conceptual shift in how physical reality itself was imagined. Faraday's picture of continuous lines of force implied that fields extend smoothly and seamlessly through space. Capturing this idea rigorously meant embracing continuity as a principle of nature, not merely a calculational convenience. Calculus provided the framework to do so, allowing forces to be treated not as isolated, discrete actions but as smoothly varying entities filling every region of space. With these tools, Maxwell would be able to transform Faraday's qualitative vision into a mathematically precise, continuous description of fields — a theory capable of exact prediction and of unifying electricity, magnetism, and light within a single, coherent structure.

The Birth of a New Language

As mathematical thought evolved in Europe toward the mid-seventeenth century, *symbolic algebra* — the systematic manipulation of symbols representing quantities — emerged as a central innovation. While ancient Greek mathematics had achieved remarkable rigor and precision, it was conceived primarily as a logical and geometric endeavor, grounded within a philosophical framework. Equations, in that tradition, were statements about relationships among quantities expressed through language and geometric construction rather than symbolic abstraction.

Algebra marked a fundamental departure from this approach. European mathematicians began to treat mathematical statements in a purely symbolic form. This separation of representation from direct interpretation created a mathematical technology: symbolic expressions could be manipulated according to fixed rules, independent of their underlying meaning. This allowed novel expressions to be generated mechanically, revealing relationships that could later be interpreted and applied. It made possible a more exploratory form of reasoning, where complex transformations could be carried out with speed and precision before their implications were fully understood.

This profound shift unfolded gradually over centuries. Recall how the ninth-century work of Muḥammad ibn Mūsā al-Khwarizmi introduced systematic methods for solving equations, providing a foundation later built upon in Europe. In the fourteenth century, Nicole Oresme anticipated the symbolic era by graphically depicting quantities changing over time. François Viète, in the late sixteenth century, went further, developing a symbolic grammar and syntax that systematized algebra into a fully formal discipline. By the early seventeenth century, René Descartes standardized this symbolic language and explicitly unified algebra with geometry. However, it was not simply the appearance of symbolic algebra, but its pioneering application to represent measurable natural laws that directly paved the way for the development of calculus in mid-seventeenth-century Europe.

Analysis of the motion of objects formed a central thread in this development. Aristotle had long maintained that sustained motion necessarily required a continuous force, consistent with common experience — objects tend to slow and stop unless continually pushed. Galileo Galilei, however, fundamentally altered this perspective through careful experimentation and rigorous reasoning. He concluded that an object in uniform motion, in principle, required no ongoing force: a ball, once set rolling along an idealized frictionless plane, would continue indefinitely at constant speed without any additional push. Force, he argued, was needed not to sustain motion, but to change an object's velocity. This striking departure from Aristotle positioned velocity as a concept of paramount importance — the primary physical quantity influenced directly by force.

The principle that uniform motion persists without force (later known as *inertia*) became the key to analyzing projectile motion, a challenge that had puzzled thinkers for centuries. While earlier figures such as Niccolò Tartaglia in the sixteenth century had correctly recognized that projectile paths were smooth curves, the precise shape remained undetermined. Galileo synthesized his insights by treating projectile flight as two simultaneous, independent motions: the horizontal motion continued steadily at constant velocity (inertia), while the vertical motion underwent constant downward acceleration due to gravity (following his principle that distance fallen relates to the square of time). These components, occurring together yet independently, produced a path whose vertical displacement was proportional to the square of its horizontal displacement. In classical geometry, this was precisely the defining property of a *parabola*, one of the conic sections studied exhaustively by Apollonius. By linking his physical principles to this established geometrical result, Galileo demonstrated that the path of an ideal projectile is parabolic — a synthesis of observation, measurement, and geometry that showed how abstract reasoning could precisely describe the natural world.

Galileo's analysis of projectile motion also underscored a deeper geometric challenge. Visually plotting the changing position of a moving object against time generally produced a curve rather than a straight line — for example, the distance–time curve for a body falling under gravity. The steepness, or slope, of such a curve at any particular point represented the object's instantaneous velocity at that moment. Likewise, determining the exact direction of travel along a parabolic trajectory required knowing the curve's orientation at each point. In geometry, this property is captured by the *tangent* line — the unique straight line that just touches the curve at the given point without cutting across it. Thus the physical task of finding an instantaneous velocity or direction of motion could be restated as the geometric task of constructing the tangent to the curve at the relevant point.

The study of tangents long predated calculus. Early Greek geometers encountered tangent problems naturally while exploring geometric figures, asking how to draw a straight line touching a circle at exactly one point. Euclid, around 300 BCE, treated tangents explicitly in the *Elements*, defining and constructing them with geometric rigor. His construction took any chosen point on a circle, drew the radius to that point, and then constructed the line through the point perpendicular to that radius. He proved that this line meets the circle at one and only one point. The proof was by contradiction: assuming an additional point of intersection led to the impossibility of an isosceles triangle containing two right angles. The absurdity of the conclusion, combined with the soundness of the logic, implied that the assumption must be false — a tangent could not possibly intersect the circle at more than one point.

As Greek geometry matured, mathematicians were drawn to more complex curves. In the third century BCE, Archimedes studied figures such as parabolas and a spiral of his own devising — the *Archimedean spiral*, traced by a point moving outward at a steady rate along a line that rotates uniformly about a fixed center. For this spiral, he defined the tangent as the straight line obtained by joining two points on the curve and then bringing those points as close together as possible, until they coincided at a single location. If a candidate line through the point cut the spiral again, however close to the point of contact, then — using the spiral's defining relationship between distance from the center and rotation — he could select another point between the contact point and the crossing point, draw a new line, and find it lay even closer to the curve's path. This refinement could be repeated without end, so no crossing line could be the tangent. The only line immune to such replacement was the one that touched the spiral at the point and lay precisely along its local course without cutting it.

In the centuries that followed, European mathematics largely preserved such classical results without major extension. In the Islamic Golden Age (eighth–fourteenth centuries), mathematicians such as Ibn al-Haytham revisited the tangent problem for curves that had received less attention in Greek works — notably the ellipse and the hyperbola. These, like the circle and parabola, were conic sections, but their geometry presented different challenges: the symmetry was more complex, the curvature varied more sharply, and the defining distance relationships required careful adaptation of existing methods. Ibn al-Haytham devised precise geometric constructions for locating the tangent line at a given point on these curves, using intersecting circles and auxiliary lines to establish a contact point that satisfied the strict non-intersecting property. Such results extended rigorous tangent work into curves important for optics and astronomy, where ellipses and hyperbolas arise naturally.

While the study of tangents was ancient, symbolic algebra was not, and by the seventeenth century there was still no general algebraic solution to the tangent problem. As the need to measure instantaneous velocities became more pressing and the use of algebra more widespread, tangents continued to be treated essentially as geometric problems — solved case by case through specific constructions, or handled by abstract reasoning of the sort Archimedes had applied to the spiral. In the early seventeenth century, Pierre de Fermat (1601–1665) began to recast this geometric reasoning in algebraic form, bridging the traditions of Archimedes and Euclid with the emerging symbolic methods of his own era.

Though Descartes is often credited with recognizing the connection between algebra and geometry, he was not the only one to do so, or even the first. As is often the case in the history of mathematics, similar insights tended to emerge from different corners, often around the same time. In fact, while Descartes was still preparing his manuscript establishing this idea, Fermat, working independently, had already developed the core principles of analytic geometry. In an unpublished manuscript titled *Introduction to Plane and Solid Loci* (Ad locos planos et solidos isagoge, circa 1636), Fermat described geometric curves as algebraic *loci* — sets of points defined by algebraic equations involving two unknown quantities. Both Fermat and Descartes drew upon the pioneering work by François Viète a few decades earlier, who had introduced letters to systematically represent unknown quantities and relationships. Fermat did not formally introduce perpendicular axes as Descartes did, or explicitly use the modern terms associated with coordinates. However, he implicitly employed a system much like modern coordinate geometry. For example, he characterized classical geometric shapes such as parabolas through conditions like "the square of the distance from each point to a given fixed line equals its distance to another fixed line", expressed algebraically as $y^2 = x$.

Fermat then went further to develop an algebraic method of tangents, built directly upon the fundamental idea that Archimedes had introduced nearly two millennia earlier in geometric form. Archimedes had shown that the tangent line to a curve could be understood as the position of a line drawn between two points on the curve as these points effectively merged into a single point. Fermat, recognizing that algebraic equations could represent curves, translated Archimedes's geometric intuition into a symbolic and algebraic procedure. His method of *adequality* (approximate equality) involved introducing an infinitesimally small symbolic term ε into the equation defining the curve, allowing him to examine the slope of a line connecting two points separated by an arbitrarily small distance — precisely mirroring Archimedes's geometric process. By doing so, he could derive two independent expressions for the "same" point: one from substituting x into the equation, and another from substituting $x + \varepsilon$. Using the algebraic expression of the curve, he could find the output, or y, for each input value, and their difference reflected how the curve changed over the tiny distance ε. By taking the ratio of these differences (difference in output over difference in input), he could compute the slope of the curve's tangent at any point.

When analyzing how one quantity varies in response to another, the concept of a "rate of change" emerges naturally. This rate describes how sensitively an output variable responds as an input variable changes over a given interval. In time, any such rate of change between variables would come to be called a *derivative*. Between two distinct points on a plane, the rate is found by taking the ratio of the vertical distance to the horizontal distance. At a single point, however, both distances are zero, resulting in the mathematically undefined form $\frac{0}{0}$. One way to avoid this is to consider intervals that become increasingly small, approaching but never reaching zero length. A term can be introduced to represent the smallest possible interval that is not quite zero — an *infinitesimal*, or infinitely small quantity. By evaluating the rate of change over these tiniest of intervals, it becomes possible to define the instantaneous rate of change at a single point without dividing by zero. This is precisely what Fermat achieved using his method of adequality.

In seventeenth-century Europe, scientific communication depended heavily on letters, as there were no scientific journals in the modern sense. Scholars and mathematicians maintained their intellectual networks through personal correspondence, exchanging ideas, results, and challenges across great distances. This loose but vibrant community was often referred to as the Republic of Letters. Central figures, notably the French mathematician and cleric Marin Mersenne, acted as hubs — receiving, circulating, and preserving correspondence among leading mathematicians, including Pierre de Fermat and René Descartes.

Initially, Fermat and Descartes exchanged letters respectfully, if somewhat cautiously. Fermat's algebraic methods for solving geometric problems — especially his tangent procedure — circulated within this scholarly network largely through Mersenne's efforts. Given the timing and nature of these exchanges, it is highly plausible that Fermat's ideas influenced Descartes as he finalized his own work on analytic geometry. Over time, as Descartes became increasingly aware of Fermat's parallel discoveries, their relationship grew strained, marked by disputes over priority and recognition.

The rivalry was notably one-sided. Fermat, a reserved magistrate who pursued mathematics as an intellectual recreation, showed little interest in claiming public credit. He freely shared his insights in correspondence, content to discuss them among a small circle of scholars. Descartes, by contrast, was deeply invested in priority, reputation, and protecting his intellectual legacy. Sensitive to criticism and quick to react when challenged, he often saw Fermat's work as a threat. The result was an asymmetrical dynamic: Descartes's competitiveness magnified the tension, while Fermat's reluctance to publish and his unassuming manner meant he was largely indifferent to formal credit. This, in part, explains why Descartes became historically dominant, widely credited with innovations that Fermat had independently discovered — and in some cases, originated first.

Historians often refer to Fermat as "the prince of amateurs", a title reflecting both the originality of his ideas and the breadth of his achievements. By profession a magistrate and counselor at the parliament of Toulouse, Fermat led a demanding legal career that consumed most of his days. He pursued mathematics as a hobby in his limited spare time — often late at night or during brief intervals of leisure. Remarkably, even with these scant opportunities, Fermat pioneered crucial ideas across a wide range of mathematical fields. Beyond his work in analytic geometry and tangents, he laid foundations in number theory, including the statement later known as *Fermat's Last Theorem*, and in probability theory, together with Blaise Pascal. In optics, he formulated the *principle of least time*, arguing that light takes the path requiring the shortest travel time between two points. This principle explained known laws of reflection and refraction — the latter having been quantified decades earlier by Willebrord Snellius (1580–1626) — and would later be shown to follow naturally from the wave theory of light developed by Augustin-Jean Fresnel (1788–1827).

However, Fermat's status as a hobbyist also meant that his mathematical reasoning was often sketchy, or incomplete by the rigorous standards of professional mathematicians. Rather than providing fully detailed proofs, Fermat frequently made brief statements or assertions in the margins of texts or in letters — such as his famous note regarding his so-called "last theorem", which claimed he had a "marvelous proof" too large to fit in the margin. Whether due to constraints on his time or because of his natural inclination toward concise intuition rather than exhaustive rigor, many of Fermat's bold insights initially seemed to be missing thorough justification. This habit of brevity often left contemporaries and later mathematicians to laboriously reconstruct or validate his methods — revealing both the brilliance and occasional incompleteness that characterized his unique approach as an amateur mathematician.

Fermat's intuitive yet informal approach appears clearly in his method for finding maxima and minima. Having already introduced his algebraic method of adequality for tangents, Fermat recognized that at points of maximum or minimum, a curve would neither increase nor decrease, reaching what he described simply as a "greatest" (*maximum*) or "least" (*minimum*) value. At such points, the tangent is horizontal and its slope is zero, meaning that an infinitesimal change in the variable should produce no change in the vertical value of the curve. Fermat added a small increment ε to the input variable, expanded and simplified the resulting expression, and then set the remaining output terms containing ε equal to zero. He applied this to algebraic equations, such as $y = x^3 - bx^2$, then discarded the ε terms on the grounds that they were infinitesimal by definition. Though Fermat himself did not offer a formal logical foundation for this critical step, he confidently presented his technique as a reliable algebraic procedure and circulated examples of it in correspondence. Many contemporaries — including Descartes and others in the Republic of Letters — expressed doubts about the rigor of this reasoning, puzzled by Fermat's willingness to omit formal justification, and leaving later mathematicians to fill the logical gap.

Yet the method proved both practical and successful for determining where the least and greatest values of an equation occur.

Fermat's technique for finding maxima and minima represented a profound conceptual step. By identifying algebraically the conditions that *extrema* satisfy, Fermat's method provided an essential tool: a systematic way to analyze the behavior of curves, optimize quantities, and explore instantaneous changes. Beyond tangents and extrema, Fermat also extended his algebraic insights to *quadrature* problems — explicitly tackling the computation of areas under curves through summations and carefully chosen comparisons. Although his quadrature methods remained somewhat informal and limited, Fermat had effectively grasped the two central concepts of calculus: the *derivative* (implicit in his tangent and extrema methods) and the *integral* (implicit in his quadrature investigations). Thus, by the time Fermat concluded his mathematical explorations, the critical conceptual insights of what would become *calculus* were all but realized. The crucial work that remained — the rigorous justification, the systematic organization, and the formalization of the connection between derivatives and integrals — would occupy mathematicians like Isaac Barrow, Isaac Newton, Gottfried Wilhelm Leibniz, and others for generations to come.

The term *quadrature* is derived from the Latin *quadratura* (squaring), a concept which can be traced back to antiquity. The ultimate goal was often stated as constructing a square having the same area as a given curved figure. Finding the area was equivalent to finding the side length of this square. Hence, finding the area became known as "squaring" or quadrature. Quadrature can be contrasted with the related concept of *rectification*, coming from the Latin *rectus* ('straight') and *facere* ('to make'). In rectification, the goal is to determine the length of a curve, which was conceived as finding the length of a straight line segment equivalent to the curve if it were "made straight".

The classical problem of determining areas enclosed by geometric curves had long fascinated thinkers. One celebrated solution was that of Archimedes, who explicitly addressed this challenge in his famous treatise, *Quadrature of the Parabola*. To determine the exact area enclosed by a parabola and a straight chord drawn across it, Archimedes employed many ingenious geometric techniques. He proved that the area of such a segment is always equal to four-thirds the area of the largest triangle that can be inscribed within it.

In one of the multiple proofs he gave, Archimedes used a process of repeated iteration. Starting with the largest inscribed triangle, he constructed two new smaller triangles in the spaces left open on either side. This process could be continued indefinitely, with the combined area of the two new triangles in each step equal to one quarter the area of the triangle from the previous step — a fact he demonstrated with careful geometric reasoning. From this, the total area of the parabolic segment could be thought of as an endless sum: first the large starting triangle, then two smaller ones making up a quarter of its area, then even smaller ones making up a quarter of that, and so on forever. Archimedes pointed out that if you set aside the first large triangle, the collection of all the remaining triangles has exactly the same structure as the original set — but every piece is one quarter the size. That means the remainder is a smaller copy of the whole. Because the first triangle and the remainder together make up the total area, and the remainder is exactly one quarter of that total, it follows that the first triangle must account for three-quarters of the total. Turning this around, the total area must be four-thirds the area of the first triangle — exactly as Archimedes claimed.

Bonaventura Cavalieri (1598–1647), an Italian mathematician and disciple of Galileo, published his well-known work on the *method of indivisibles* in 1635, two years before Descartes's text on geometry. Cavalieri's contributions are perhaps best understood as a rediscovery and popularization of intuitive ideas articulated by Archimedes generations earlier. Although Cavalieri is said to have had no direct access to Archimedes's treatise *The Method*, his core insight paralleled Archimedes's private heuristic: envisioning geometric figures as composed of infinitely thin slices, or *indivisibles*, whose areas could be directly compared. Archimedes had confined this to heuristic reasoning to be justified later by the method of exhaustion; Cavalieri instead openly promoted it as a usable mathematical tool. His real innovation was not so much a new discovery as a powerful act of dissemination — bringing into wide mathematical discussion an Archimedean idea that had existed implicitly but had never been broadly embraced.

Cavalieri's approach met with sharp criticism from some contemporaries. The most vocal was the Swiss Jewish Jesuit mathematician and astronomer Paul Guldin (1577–1643). Guldin's objections reflected a broader conservative reaction rooted in classical Greek geometry and Aristotelian philosophy, often championed by the Jesuit order. His central charge was the *dimensionality problem*: that a two-dimensional area could not be composed of one-dimensional lines (which have zero area), nor a three-dimensional volume of two-dimensional planes (which have zero volume). Summing infinitely many zero-magnitude elements, he argued, could not produce a finite magnitude. This violated the traditional understanding of dimensions. Guldin also questioned whether a continuous magnitude — a line, area, or volume — could truly be made up of discrete indivisibles. Aristotelian doctrine held that continua were infinitely divisible but not composed of ultimate indivisible parts. To many, Cavalieri's method seemed to import a kind of geometric atomism, which was philosophically suspect. Moreover, it appeared inferior to Archimedes's method of exhaustion, which rigorously trapped a magnitude between inscribed and circumscribed figures and showed that the difference could be made arbitrarily small, avoiding any direct appeal to indivisibles. To Guldin, Cavalieri's approach was a non-rigorous shortcut that sidestepped the careful reasoning of the exhaustion method.

In later editions of his work, Cavalieri responded to these criticisms. He argued that the method of indivisibles was essentially a faster version of the method of exhaustion, giving the same results with less labor, and tried to demonstrate their equivalence in some cases. To address the dimensionality problem, he sometimes suggested picturing the "lines" making up an area as very narrow rectangles, and the "planes" making up a volume as very thin slabs — giving the indivisibles a tiny thickness. He stressed that the method consistently produced correct results, which could often be checked by other means. Still, his defenses did not satisfy those who demanded strict Euclidean rigor, as they left unresolved the fundamental logical objections about summing elements of zero magnitude or the nature of the continuum.

Another critic was Gilles Personne de Roberval (1602–1675), a leading French mathematician devoted to the standards of classical geometry. Working in Paris, Roberval questioned Cavalieri's reliance on indivisibles, citing their initial vagueness about dimensionality and their lack of explicit geometric construction. Although Cavalieri eventually refined his method by assigning indivisibles an infinitesimal thickness, addressing some of Roberval's concerns, Roberval himself remained committed to explicit geometric precision. This rigor is evident in his celebrated solution to the problem of finding the exact area under one arch of the *cycloid* — the curve traced by a point on the rim of a circular wheel as it rolls along a straight line.

The cycloid had attracted attention since Galileo began studying it around 1599. Galileo was drawn to it because it was a novel curve generated by motion — unlike the conic sections studied since antiquity — and posed intriguing challenges, such as determining its area. He even considered the inverted cycloid arch as a potential bridge design, though it was never adopted. Galileo attempted to measure the area experimentally by making physical models and weighing them, finding a result close to three times the area of the generating circle, though uncertain if it was exact.

By the 1630s, understanding the cycloid had become a prominent challenge for European mathematicians — both a practical puzzle for engineers and a theoretical test for geometers. Between about 1634 and 1640, Roberval "solved" it using purely classical geometry. He enclosed a cycloidal arch within a rectangle whose width was the circumference of the generating circle and whose height was the circle's diameter, so the rectangle's area was four times that of the circle. He then showed, by finite rearrangements and explicit comparisons, that the area left between the cycloid and the rectangle was exactly equal to the area of the generating circle. Thus the area under the cycloid was exactly three times the circle's area.

By the mid-seventeenth century, the cycloid's appeal had grown beyond its geometric elegance to include its surprising mechanical properties. These proved especially important in clockmaking, a field critical to navigation, astronomy, and experimental physics. The most significant advance linking the cycloid to accurate timekeeping came from the Dutch engineer Christiaan Huygens (1629–1695), whose work connected geometric curves directly with mechanical motion.

Huygens's critical insight arose around 1657, as he sought to address a fundamental problem with pendulum clocks. A simple pendulum swinging along a circular arc does not keep exactly the same period for different amplitudes; as its swing diminishes, small variations in period accumulate into noticeable errors. Huygens realized that if the pendulum bob were made to follow a cycloidal path instead of a circular one, the swing would be perfectly *isochronous* — the period would remain constant regardless of amplitude. In other words, no matter how high the bob is released along the path, it reaches the bottom in exactly the same time.

To achieve this in practice, he constrained the pendulum's motion with specially shaped guides, or *cycloidal cheeks*, which forced the string to wrap and unwrap along an inverted cycloid so the bob followed the correct curve. He then proved the result mathematically in his 1673 treatise *The Pendulum Clock* (Horologium Oscillatorium), using rigorous classical geometry. His proof showed that the cycloidal path precisely offsets the variation in gravitational acceleration along different parts of the swing, giving a uniform period for any amplitude.

Huygens's reasoning combined two main ideas. First, for a curve to have the isochronous property, the acceleration toward the lowest point must be proportional to the displacement from that point. This ensures that the extra distance traveled in a large swing is matched exactly by the increase in average speed. Second, on a cycloid, the displacement from the bottom is proportional to the ratio of the bob's vertical height to its straight-line distance from the bottom. Because gravitational acceleration is also proportional to this ratio, the condition for isochronism is satisfied. Huygens established these properties through geometric arguments in the spirit of Archimedes, enriched with the infinitesimal techniques developed by Roberval and others.

But simply knowing the ideal path was not enough. Huygens had to find a way to make the pendulum follow it exactly. In doing so, he discovered an extraordinary geometric fact: when a string unwraps tangentially from a cycloidal cheek, the bob's path is itself an identical cycloid. This "self-generating" property is exceptionally rare in geometry, and in this case almost miraculously convenient — the very curve required for perfect timekeeping also serves as the template that mechanically produces it. This perfect harmony between mathematical form and physical function bridged the gap between theory and working mechanism.

The amount of mathematical and engineering insight represented by the development of Huygens's improved pendulum clock was enormous. It took decades to build upon the revelations gleaned from Galileo's work, to understand the geometry of the cycloid and its *rectification* (its equivalent straight-line distance), and finally to recognize the isochrone property and realize it in a functioning clock.

Whether this influenced Huygens's personal faith is hard to say from his technical writings, but in the broader seventeenth-century culture of natural theology, such harmony between mathematics and nature was often read as evidence of divine wisdom. Many saw the deep rational order revealed by geometry and mechanics not as a coincidence, but as a sign that the laws of nature were both intelligible and imbued with elegance by design.

The cycloidal pendulum clock represented a monumental leap in accuracy. It quickly surpassed all previous designs and had an immediate impact on navigation by improving the determination of longitude at sea — a long-standing challenge for maritime powers. Latitude could be found fairly easily by measuring the sun's noon altitude or the elevation of known stars, but longitude required knowing the precise time difference between two locations. A ship could carry a clock set to the time at its home port; each day, the navigator would determine local noon by observing the sun, then compare it to the time on the ship's clock. Every hour of difference corresponded to fifteen degrees of longitude (three-hundred-sixty degrees divided by twenty-four hours). Adding or subtracting this from the known longitude of the home port yielded the ship's position east or west. While Huygens's pendulum clock greatly improved this method, it was not a complete solution for use at sea. Pendulum clocks are sensitive to motion and changes in gravity, and work best when stationary and level. On a rolling ship, their accuracy still suffered. Later, spring-driven *marine chronometers* overcame these limitations, but the cycloidal pendulum was a decisive step forward.

Beyond pendulum dynamics, Huygens quantified centrifugal force, providing a foundation for analyzing rotational motion. In optics, he formulated *Huygens's Principle*, which described each point on a wavefront as a source of secondary wavelets that spread in all directions, explaining reflection, refraction, diffraction, and interference. He discovered Saturn's moon Titan and correctly interpreted the structure of its rings.

These achievements exemplified a style of work that was becoming increasingly uncommon. Huygens relied on rigorous, case-specific geometric proofs, each crafted with bespoke ingenuity for the problem at hand. In his hands, mathematics was still an art of tailored solutions. But around him, a different mathematical culture was taking shape — one built on symbolic algebra and calculus, emerging as general-purpose "technologies" that could be taught, learned, and applied broadly across disciplines. Huygens's cycloidal pendulum, for all its brilliance, was both a pinnacle of the traditional style and a marker of the transition toward these more systematic methods.

Blaise Pascal (1623–1662) shared with Huygens a deep interest in intricate mathematical problems such as those posed by the cycloid, as well as in foundational questions that led to the birth of probability theory. Like Huygens, Pascal was fascinated by both the elegant geometry and the practical implications of the cycloid. Both recognized that solving complex geometric challenges related to this curve could yield insights extending beyond pure mathematics, shaping the emerging fields of physics and mechanics. Similarly, Pascal's engagement with probability paralleled Huygens's later work; each sought to mathematically capture uncertainty, rationalize decision-making, and uncover the hidden logic governing games of chance and real-world outcomes.

At just sixteen, Pascal wrote his *Essay on Conics* (1640), introducing what would become known as *Pascal's theorem*. The theorem states that for a hexagon inscribed in any conic section, the intersections of three pairs of opposite sides always lie on a single straight line. This result laid foundational ideas for the modern field of *projective geometry*, which studies geometric properties preserved under *perspective transformations*, rather than traditional measurements like lengths and angles. Pascal also constructed a mechanical calculator, the Pascaline (1642), motivated by the practical need to assist his father with complex tax computations.

Pascal's work was marked by a rigorous pursuit of clarity and precision. In hydrostatics and fluid mechanics, his principle known as *Pascal's Law* arose from careful experiments and deduction. He reasoned that pressure applied to an enclosed fluid is transmitted undiminished throughout the fluid and acts equally in all directions. From thought experiments, he concluded that this principle must apply universally, regardless of the container's shape or orientation. This conceptual leap became foundational for understanding fluid dynamics and enabled the design of hydraulic machinery.

Pascal's intellectual rigor found perhaps its most innovative expression in the emerging discipline of probability theory. Around 1654, he became intrigued by questions posed by gamblers, particularly the *problem of points* — how to fairly divide the stakes if a game ends prematurely. Corresponding with Pierre de Fermat, Pascal dissected the logic of such chance-driven scenarios. His breakthrough lay in recognizing that randomness obeys consistent rules. By clearly defining all possible outcomes and analyzing their relative frequencies, he showed that uncertain future events could be quantified and managed rationally. These insights laid the groundwork for modern probability theory, influencing decision-making in economics, social sciences, and moral philosophy.

In June 1658, Pascal publicly issued an elaborate set of cycloid challenges under the pseudonym *Amos Dettonville*. He posed problems such as determining the exact area beneath a cycloid arch, the length of its arcs, and the volumes generated by rotating cycloidal segments about a line — essentially *proto-calculus* exercises. While the pseudonym was unlikely to truly conceal his identity in the small, interconnected mathematical community, it created a forum for impartial competition, blending seriousness with playful intellectual intrigue. Pascal supplied elegant solutions of his own, using sophisticated geometric constructions that subtly employed ideas akin to *limits* and *infinitesimals*, anticipating techniques that would later become part of formal calculus.

Pascal's contributions also included his influential study of what is known as *Pascal's Triangle*. Although triangular number arrays were known to Indian, Persian, and Chinese mathematicians centuries earlier, they were mainly regarded as numerical curiosities. In his *Treatise on the Arithmetical Triangle* (Traité du Triangle Arithmétique, circa 1654), Pascal explicitly connected the triangle's structure to symbolic algebra, particularly the expansion of terms from products of sums. By interpreting each number as a combination — the number of ways to choose objects from a set — he revealed the triangle's deep combinatorial and algebraic significance. It became a practical computational tool for symbolic algebra. Pascal likely did not know of the triangle's ancient origins, independently rediscovering and greatly enriching its mathematical interpretation.

Operating contemporaneously with Huygens and following closely after Pascal, René-François de Sluse (1622–1685) distinguished himself through an algebraic generalization of tangent methods. Between about 1655 and 1659, Sluse developed an explicitly algebraic procedure — known as *Sluse's Rule* — for finding tangent lines to curves defined by certain classes of equations, streamlining the more elaborate geometric and infinitesimal approaches in use at the time.

Sluse's work was closest in spirit to Fermat's method of tangents and extrema. Both aimed to produce general rules for determining slopes from equations, but where Fermat's procedure often blended algebra with curve-specific geometric reasoning, Sluse reformulated the process as a compact symbolic algorithm. His rule could be applied directly to the equation of a curve to yield the tangent slope, without constructing auxiliary figures or working through proportional arguments.

While Pascal's reputation rested on *combinatorics*, probability, and geometric ingenuity, and Huygens pursued curve problems through explicit geometric and physical constructions, Sluse's approach offered a more concise, purely algebraic pathway. Huygens, in correspondence with Sluse in the late 1650s and early 1660s, adopted the method in his own work, integrating it into his geometric analyses to increase both their speed and generality.

John Wallis (1616–1703), an English acquaintance and admirer of Fermat, systematically explored the power of algebraic methods to handle infinite processes. Like Fermat, his work often fell short of modern standards of rigor, but his 1655 *Arithmetic of Infinites* (Arithmetica Infinitorum) was a landmark. In it, Wallis boldly manipulated infinite series, summations, and products — most famously expressing the transcendental number π as an infinite product of rational numbers, known as the *Wallis Product*. Drawing on the geometric intuitions of Cavalieri and Pascal, he translated them into the language of algebra, producing a more systematic approach to finding areas under curves. He extended quadrature methods to curves defined by fractional and negative powers — such as $y = x^{\frac{1}{2}}$ and $y = x^{-1}$ — a crucial step toward a general theory of *integration*. Wallis also introduced the ∞ symbol as a finite representation of the notion of infinity.

Where many contemporaries still let geometry lead, Wallis increasingly allowed symbolic algebra to dictate the path forward. After calculating the areas under curves for specific integer exponents like x^2 and x^3 and discerning a clear pattern, he encapsulated it in a general algebraic rule: the area under $y = x^k$ is equal to $\frac{1}{k+1}$. His crucial move was to trust the algebraic form itself, extrapolating its validity to exponents far beyond those from which it had been derived — including fractions and negative numbers. This was like learning a recipe that calls for "1 cup liquid" and finding it works with water or milk, then assuming it will work equally well with oil or honey — trusting the symbolic quantity over the very different properties of each ingredient. That confidence in the universality suggested by the algebraic expression allowed him to unify the quadrature of a vast range of curves, but it also made his methods controversial to those demanding step-by-step deductive proof.

As Europeans began grappling with the idea of symbolically manipulating finite, infinite, and infinitesimal quantities, they also confronted the philosophical implications of this mathematical technology. The difficulties came to a head in a decades-long debate, sparked in part by an example introduced by Evangelista Torricelli (1608–1647), a student of Galileo. Around 1641, Torricelli described a curious solid, later called *Torricelli's Trumpet*, which combined a finite volume with an infinite surface area. It was formed by rotating the hyperbola $y = \frac{1}{x}$ about the "x-axis" from $x = 1$ to ∞, producing a graceful, tapering horn that drew ever closer to the axis as it extended without bound. The image of a vessel that could be filled with paint but never painted over became a potent symbol in disputes over the legitimacy of infinitesimal methods.

The power of these methods bred both excitement and unease. Their effectiveness was often stunning, yet they seemed built on shifting sands. *Infinitesimals* — quantities smaller than any assignable magnitude, but somehow not quite zero — defied precise definition within established logical frameworks. Operations involving them could appear paradoxical, like dividing by zero or claiming that adding infinitely many "nothings" might produce "something". Some feared that the legendary certainty of mathematics, its status as the most rigorous of human pursuits, was under threat. This unease found one of its most forceful voices in Thomas Hobbes (1588–1679). Already celebrated — and feared — for the unflinching logic of *Leviathan*, Hobbes had come to revere the deductive structure of Euclid's *Elements* late in life, seeing in it the pinnacle of human reasoning: a system built from clear definitions and irrefutable proofs. He distrusted algebraic symbols, deriding them as "a scab of symbols" that severed mathematics from the intuitive, visual clarity of geometric demonstration. To him, counter-intuitive results like *Torricelli's Trumpet* — a shape with finite volume yet infinite surface — were not marvels, but evidence of the absurdity of the methods that produced them. For Hobbes, this was a battle to preserve the foundations of certainty in knowledge.

Hobbes's challenge was aimed squarely at John Wallis, a figure of immense mathematical talent and institutional stature. Appointed Savilian Professor of Geometry at Oxford in 1649 — one of two prestigious chairs in mathematics and astronomy endowed by Henry Savile (1549-1622) in 1619 — Wallis occupied a central position in English academic mathematics. His reputation extended well beyond the university: during the English Civil War he had gained renown as a cryptographer, successfully deciphering Royalist ciphers for the Parliamentarian side.

The sustained and public nature of the Hobbes–Wallis combat was made possible by the print revolution. Before the press, scholarly exchange relied on scarce, hand-copied manuscripts, restricting complex debate to small institutional elites and making rapid, widespread argument nearly impossible. By the mid-seventeenth century, printing on increasingly affordable paper allowed books and pamphlets to be produced in volume and distributed through expanding networks of printers and booksellers. The two men's attacks and counterattacks could circulate quickly, reaching a wide — though still primarily scholarly — audience.

England's intellectual landscape was also being reshaped by institutions. Informal meetings of "natural philosophers" in London and Oxford during the unsettled 1640s and 1650s culminated in the formal founding of the Royal Society of London after a meeting on November 28, 1660, with Royal Charters in 1662 and 1663. Figures such as John Wilkins, Robert Boyle, Christopher Wren, Robert Hooke, and Wallis himself were central to its creation. The Society's explicit aim — "Improving Natural Knowledge" — and its motto "Nullius in verba" (Take nobody's word for it) reflected a commitment to collaborative experiment and observation. While the Society never formally intervened in the Hobbes–Wallis dispute, it shaped the arena in which it played out. As a founding Fellow, Wallis enjoyed an institutional base, a network of allies, and a platform that reinforced his scientific standing. Hobbes, deeply critical of Boyle's experimental philosophy and the Society's approach to knowledge, remained an outsider. His exclusion from this increasingly influential body framed his mathematical critique as resisting the current represented by Wallis and the Society, deepening his isolation within England's dominant scientific circles.

The clash became most personal and mathematically concrete in their long-running battle over the classical problem of *squaring the circle*. Hobbes, convinced of the sufficiency of ancient geometry, insisted he could construct — using only straightedge and compass — a square equal in area to a given circle, a feat that many geometers of the time already suspected could not be achieved. From the 1650s onward he published intricate constructions claiming success. Wallis, with relish and formidable skill, dismantled each in print, exposing flawed deductions and incorrect geometric assumptions. These were failures even by the Euclidean standards Hobbes championed. For Wallis, publicizing such errors was strategically potent: it undercut Hobbes's authority to criticize the methods of professional mathematicians when he could not master the geometry he idolized. Broadcast through print, this specific quarrel added a layer of public humiliation to Hobbes's position.

While the venomous feud was fundamentally an English affair, it resonated across the European Republic of Letters. Scholars abroad acted as spectators and, at times, participants. In France, Antoine Arnauld (1612–1694) and Ignace Gaston Pardies (1636–1673) took aim at Hobbes's work. Arnauld attacked the foundations of Hobbes's philosophy — his materialism, determinism, and perceived logical flaws — arguing that such errors inevitably undermined his reasoning in every field, including mathematics. Pardies examined Hobbes's scientific writings, pointing out specific mistakes in his mechanics and geometric proofs, and in doing so gave continental validation to the English view that Hobbes's technical work was unsound.

Privately, some of Europe's most eminent mathematicians reinforced this verdict. Christiaan Huygens, widely regarded as one of the foremost scientific minds of the age, agreed that Hobbes's mathematical efforts were fundamentally flawed, even as he occasionally voiced reservations about Wallis's combative style and insisted on the highest rigor in all argument. Gilles Personne de Roberval, famed for his uncompromising geometric exactness, dismissed Hobbes's claims outright. These informed rejections, shared within elite correspondence networks, carried immense weight and confirmed Hobbes's isolation from the scientific mainstream.

Hobbes, by contrast, was deeply skeptical of the methods and collective judgment emerging from circles such as Wadham College, Oxford, and later the Royal Society. He preferred a solitary, deductive approach built on first principles, and distrusted knowledge derived from contrived experiments or the consensus of a group. Ironically, while championing rigor, he vigorously defended his own mathematical work even when it contained what many regarded as clear logical flaws. Convinced that his critics either misunderstood his core ideas or relied on techniques — such as infinitesimals — that he believed lacked a secure foundation, he would not concede error. As a result, although his specific scientific claims were widely rejected, Hobbes's relentless — if arguably misguided — insistence on strict demonstration unintentionally highlighted the urgent need for stronger logical foundations for the algebraic and infinitesimal methods his contemporaries were advancing.

Meanwhile, Isaac Barrow (1630–1677), Isaac Newton's predecessor at Cambridge, made contributions that directly paved the way for the formal development of calculus. While John Wallis had focused on an algebraic approach to infinitesimals, Barrow pursued a more geometric route, building upon the work of Fermat and drawing inspiration from ancient Greek methods, particularly those of Archimedes. In his *Geometrical Lectures* (Lectiones Geometricae, 1670), Barrow described a method for finding tangents to curves that came very close to modern differentiation. He used what is called a *differential triangle* — a tiny right-angled triangle formed by minute increments of the horizontal and vertical coordinates, and a small segment of the curve representing the slope (or tangent) at an infinitesimal scale.

Barrow's most important advance over prior methods was recognizing and explicitly stating a version of the *Fundamental Theorem of Calculus*: the inverse relationship between differentiation and integration. He showed that finding the tangent to a curve (*differentiation*) and finding the area under it (*integration*) could be seen as opposite processes. His work, however, remained firmly rooted in geometry; he did not yet develop the symbolic, algebraic machinery that would make calculus a more general computational tool.

Barrow's student, Isaac Newton (1643–1727), took his teacher's profound geometric understanding of the inverse relationship between tangents and areas and elevated it into a systematic framework. In his manuscript *Method of Fluxions and Infinite Series* (Methodus Fluxionum et Serierum Infinitarum, circa 1671), Newton introduced the terms *fluxions* (rates of change, analogous to derivatives) and *fluents* (quantities that change, analogous to integrals). Within this system, he provided formal, generalized proofs of the "fundamental theorem". Newton, however, never published this work during his lifetime, and his specific system is not the genesis of modern calculus notation.

The term *calculus* originates from Latin, literally meaning a 'small stone' or 'pebble'. The word reflected the ancient practice of using pebbles for counting and performing arithmetic, often on an abacus or counting board, and in classical Latin it came to signify computation, reckoning, or any method of calculation. In this broader sense — as a general term for a system of computation or reasoning — it was in use long before it acquired a specific mathematical meaning. That transition occurred primarily through the work of the German philosopher and mathematician Gottfried Wilhelm Leibniz (1646–1716). As Leibniz independently developed his systematic methods for dealing with rates of change and accumulation, in parallel with Newton's *Method of Fluxions*, he named them *calculus differentialis* (calculus of differences) and *calculus integralis* (calculus of sums). He chose the word *calculus* deliberately, to signal that these were general systems of computation applicable to continuous quantities. Because his terminology and concise notation were published early and adopted widely, especially across continental Europe, *calculus* became the standard name for the entire branch of mathematics that grew from this work.

Leibniz aimed to create a universal symbolic method for solving geometric problems such as finding tangents and areas. His most important insights related to the *Fundamental Theorem of Calculus* appeared in two journal papers. In *A New Method for Maxima and Minima...* (Nova Methodus pro Maximis et Minimis..., 1684), he introduced his differential calculus, along with the notation dx and dy for infinitesimals and explicit rules for differentiation. In *On a Hidden Geometry and the Analysis of Indivisibles and Infinites* (De Geometria Recondita et Analysi Indivisibilium atque Infinitorum, 1686), he introduced his integral calculus and demonstrated the fundamental principle. Here he presented the integral sign $\int$ — an elongated 'S' for summa, the Latin word for 'sum' — and showed that integration ($\int y\,dx$, summing infinitesimal rectangles to find area) is the inverse of differentiation. He expressed this relationship as $\frac{d}{dx}\int y\,dx = y$, making it visually clear and algebraically manifest. His conception of $\int dy$ as the sum of infinitesimal differences, equalling the total difference in y, offered a powerful heuristic understanding of the process.

The rapid and widespread adoption of Leibniz's methods and notation, compared to Newton's, can be traced to his publication strategy, communication network, and the influence of key allies. Leibniz published his results promptly, ensuring their early visibility across Europe, and actively shared his notation and methods in correspondence with mathematicians such as the Swiss brothers Jakob Bernoulli (1654–1705) and Johann Bernoulli (1667–1748) in Basel. The Bernoullis became both prolific developers and tireless promoters of Leibnizian calculus through their own research and teaching. A decisive step came when Johann Bernoulli tutored the French marquis Guillaume de L'Hôpital (1661–1704), leading to L'Hôpital's publication of the first calculus textbook, *Analysis of the Infinitely Small for the Understanding of Curves* (Analyse des Infiniment Petits pour l'Intelligence des Lignes Courbes, 1696). Based entirely on Johann Bernoulli's lectures and using Leibniz's notation exclusively, this text became the standard throughout continental Europe, entrenching the Leibnizian framework for generations of students and researchers. In contrast, Newton was reluctant to publish his *Method of Fluxions*, which did not appear until 1736, after his death. The absence of an early, accessible Newtonian textbook comparable to L'Hôpital's meant that the Leibnizian system, promoted vigorously by influential figures and formalized in the first printed text, spread far more rapidly and broadly.

Much the way an idea gradually forms in one's mind, pieced together from hints and intuitions before it is finally crystallized and spoken by the lips, the core concepts underpinning calculus circulated for hundreds — if not thousands — of years before finding their full expression. While ancient mathematicians such as Archimedes performed remarkable feats in calculating areas and volumes, they lacked a universal, systematic method linking these interior properties directly to boundary behavior through the concepts of rates of change and accumulation. They did not yet have the machinery to express this as a general principle. What the Greeks were essentially missing — and what calculus uniquely provided — was a method connecting cumulative quantities within a region (such as area or volume) to the behavior of a related quantity on its boundary.

Through the *Fundamental Theorem of Calculus* and later generalizations such as *Stokes's Theorem*, this principle took precise form: the sum (*integral*) of a rate of change (*derivative*) over a region is completely determined by the net value of the original quantity on the boundary. An area or volume, viewed as the integral of infinitesimal elements, can thus be calculated by evaluating a related integral solely along the edge or surface. Even after Barrow, Newton, and Leibniz, it would take many years before this core idea could be stated as a sharp truth.

For Newton, the catalyst for his most intense period of mathematical creation was, unexpectedly, a disaster: the Great Plague of London. In mid-1665, shortly after graduating from Cambridge, he fled the outbreak and retreated to his family home at Woolsthorpe Manor, remaining there until early 1667 — a span of about twenty months. This period, often called his *Annus Mirabilis*, or 'Year of Wonders', produced an astonishing concentration of breakthroughs. During these months, he laid the foundations of his version of differential and integral calculus, as evidenced by surviving notebooks and later accounts. Alongside this, he developed his *generalized binomial theorem* — essential for expanding expressions such as $(x+y)^r$ into infinite series — made decisive discoveries in optics (demonstrating that white light is composed of all colors), and began formulating the law of universal gravitation. The sheer density of these breakthroughs in such a short time speaks to his genius, operating largely independently, and driven by an intense desire to uncover nature's mathematical underpinnings.

When Cambridge reopened in spring 1667, Newton returned with the clear aim of pursuing an academic career. The standard path required securing a Fellowship at his college — a competitive position awarded by internal election on the basis of demonstrated merit and strong recommendations. Such posts provided financial support and freedom for advanced study toward a Master's degree and independent research. With an exceptional undergraduate record and the support of his tutor, Isaac Barrow, Newton was elected a Fellow in 1667 and received his MA in 1668.

By 1669, he had consolidated some of his mathematical breakthroughs, particularly his use of infinite series to calculate areas, into the manuscript *On Analysis by Equations with Infinitely Many Terms* (De Analysi per Aequationes Numero Terminorum Infinitas, circa 1669). Though unpublished, Newton shared it with Barrow, who passed it to John Collins (1625–1683), a prolific mathematical correspondent who distributed manuscripts and scientific news through an extensive letter network. Collins's limited circulation of De Analysi helped establish Newton's reputation for exceptional ability among influential mathematicians.

In October 1669, Barrow — only thirty-nine years old but already a highly respected Lucasian Professor of Mathematics — resigned his prestigious chair to focus on theology. He is generally understood to have done so with the specific aim of clearing the way for Newton's appointment. Barrow's influence with the electors, combined with Newton's demonstrable brilliance, outweighed concerns about Newton's youth and reserved temperament. At just twenty-six, Newton assumed the Lucasian Professorship, gaining the platform and security for his later scientific career. Even so, he kept his more systematic calculus treatise, the *Method of Fluxions* — composed around 1671 as a formal account of his 1665–1667 work — private for decades.

Prior to Newton, mathematicians had developed methods to find tangents (essentially differentiation) and perform quadrature (essentially integration) for certain types of algebraic expressions. These techniques worked reliably mainly for *polynomials* — expressions such as $ax^k + bx^{k-1} + ...$ involving variables raised to non-negative whole-number powers. Fermat had algebraic techniques for tangents and for finding extrema of polynomials, while Cavalieri, Fermat, Wallis, and others found ways to compute areas under curves such as $y = ax^k$ for arbitrary values of k. For more complex expressions, however, methods were often limited, relying on specific geometric tricks or ad-hoc approaches.

Installed as Lucasian Professor, Newton turned immediately to advancing the mathematical capability he had inherited from his predecessors. He dramatically expanded this capability by introducing a systematic approach grounded in his mastery of infinite series — representing complex expressions as infinite sums such as $a_0 + a_1 x + a_2 x^2 + \ldots$. Using his *generalized binomial theorem*, he could rewrite a vast range of equations as power series (infinite series arranged by powers of x). This transformation was key: once in series form, differentiation or integration could be performed simply by applying the rules term by term in the sum. Newton also extended his framework beyond purely algebraic forms, incorporating known results and geometric methods to handle *transcendental* expressions — non-arithmetic expressions such as logarithms and trigonometric ratios. He clearly articulated and applied systematic rules for differentiation, including sums, products, quotients, and powers, as well as the chain rule for composite expressions. By enlarging the class of equations that could be analyzed (especially through series expansion) and by establishing consistent, general rules for working with them, Newton created a powerful and comprehensive calculus.

A key aspect emerging in clarity from Newton's framework is the difference in complexity between the two fundamental operations of calculus. Differentiation is an almost rote procedure for finding instantaneous rates of change. Although the core insight of calculus reveals that integration is the inverse operation, there is a stark asymmetry at play. While finding rates can often be done algorithmically by following step-by-step rules, the inverse process of calculating accumulation has no such universal procedure. Finding an integral often demands a cognitive leap — recognizing the underlying expression as the derivative of a known form, or the insight to apply an ingenious transformation. Newton not only systematized the rules for the more mechanical process of differentiation, but also provided a powerful technique via infinite sums to bridge the gap when this cognitive leap for direct integration is not apparent, enabling the inverse to be calculated even in many complex cases.

Newton's first significant step onto the public scientific stage was with his work on optics. In 1672, he sent his paper, *New Theory about Light and Colors*, to the Royal Society, detailing his prism experiments and his conclusion that white light is a composite mixture of immutable colored rays, each with its own ability to refract. Newton's experiments decisively showed that when a single color isolated from a prism's spectrum was passed through a second prism, its color remained unchanged and it was refracted by a unique, characteristic angle. He also showed that the dispersed colors from the prism could be recombined into white light by either a lens or another inverted prism. His paper was not simply published; rather, according to the Society's practice — which served as the primary mechanism for scientific discourse and peer review at the time — Newton's letter was read aloud by the Secretary during a meeting. The Secretary at this time was Henry Oldenburg (1619–1677), a German natural philosopher who was pivotal in facilitating scientific communication across Europe, and the reading immediately sparked intense debate.

Robert Hooke (1635–1703) was at this time the highly influential Curator of Experiments for the Royal Society, and his wide-ranging work included hypotheses on light and color published in *Micrographia* (1665). He praised Newton's experiments but fiercely attacked the novelty and validity of Newton's theory, claiming that his own earlier ideas about light modification provided a better explanation and that Newton had not sufficiently acknowledged them. Newton defended his work vigorously, emphasizing the mathematical precision his theory offered, which Hooke's lacked, leading to several rounds of heated exchanges via letters read at Society meetings. Newton maintained that his direct experiments and mathematical model proved a fundamentally different theory, and that Hooke's qualitative ideas were irrelevant to his conclusions.

At the same time, Christiaan Huygens engaged Newton (via Oldenburg) in significant debate. Respectful but unconvinced, Huygens objected on theoretical grounds, preferring his own wave-based explanation. He found it difficult to accept Newton's view that white light was a composite of immutable colored rays, instead seeing white light as simpler, with colors arising from modification. He also argued that Newton's model could not adequately explain phenomena such as diffraction (light bending around obstacles) or double refraction (the splitting of light in crystals). That Newton — still early in his public career — did not simply welcome such engagement from a figure of Huygens's stature was due in part to his personality: extremely sensitive to criticism and deeply convinced of his own results, he often found public debate taxing, regardless of tone.

Sustained challenges from both Hooke and Huygens consumed considerable time and energy. The acrimonious back-and-forth with Hooke dragged on for years, and Newton's letters to Oldenburg from 1672–1676 reveal his growing frustration with what he called the "litigious" nature of public scientific discourse. More than once he threatened to withdraw from it altogether — and by 1676, he largely had. Though fulfilling his professorial duties with lectures (reportedly sparsely attended, perhaps due to their advanced content or Newton's delivery), he devoted most of his energies through the late 1670s and early 1680s to private research in alchemy and theology.

His mathematical work did continue, as shown in his 1676 correspondence with Gottfried Wilhelm Leibniz. Leibniz, writing to Oldenburg from Paris, requested information on recent English work on infinite series and methods for tangents and quadratures — topics he was himself rapidly advancing. Oldenburg sought

Newton's input, and Newton replied with two letters that listed many results but concealed his core fluxional methods in anagrams — coded rearrangements of letters — to secure priority without revealing the techniques. One, when decoded, read: "Given an equation involving any number of fluent quantities, to find the fluxions"; and vice versa. Leibniz responded by describing his own differential method and results, signalling that he was not merely a student seeking instruction but an active researcher with methods of his own. He was impressed by the scope of Newton's results — particularly the generality of the binomial theorem and the variety of series expansions — and recognized in them the same centrality of infinite series he was exploring himself. His reaction to the anagrams, however, was one of frustration: he acknowledged they likely concealed important discoveries, but lamented that such secrecy hindered collective progress. He saw it as a priority-staking maneuver, but felt confident in his own, different approach, which he was willing to communicate more openly — though he, too, withheld full publication at this stage.

Oldenburg's sudden death in September 1677 deprived Newton of his primary conduit to the European scientific community. In the years that followed, Newton's focus shifted even more to his all-consuming private studies in alchemy and theology. He filled notebooks with alchemical experiments and exegetical work that led him to a profoundly heterodox conclusion: a rejection of the Trinity as a later corruption of Christianity. This anti-Trinitarian conviction, kept intensely secret throughout his life, carried a sharp irony given his position at Trinity College; its disclosure would have been disastrous to his career and standing. These experiences reinforced his caution — perhaps bordering on paranoia — about revealing work in progress, especially on matters that might provoke controversy or disputes over credit.

In late 1677, Robert Hooke succeeded Oldenburg as one of the Royal Society's Secretaries (serving until 1682). In November 1679, despite their earlier conflict over optics, Hooke initiated correspondence with Newton, expressly to resume philosophical discussion. In it, he outlined an approach to planetary motion that combined tangential velocity with an inward central attraction, and crucially proposed that this attraction varied inversely as the square of the distance — asking Newton's opinion on the resulting trajectory. Hooke's motives were likely mixed: fulfilling his official role, genuine interest in a central scientific problem, and perhaps a lingering desire to engage or compete with Newton. Newton replied, discussed the concepts, and even sent a diagram — though it contained an error, and he withheld any complete mathematical derivation of the orbit. Their history, coupled with the sensitivity of priority in such a fundamental discovery, almost certainly made Newton reluctant to reveal his proof before he was ready to secure full credit.

The problem of deriving planetary orbits from the inverse-square law remained unsolved in public view. By early 1684, the question was being discussed in London by

Edmond Halley (1656–1742), Robert Hooke, and Christopher Wren (1632–1723), the polymath professor of astronomy at Oxford, later famed as architect of Saint Paul's Cathedral. Hooke again asserted that he had the solution, but when pressed, he repeatedly failed to produce a mathematical proof (Wren had even offered a prize). Recognizing Hooke's inability to deliver and knowing Newton's formidable mathematical reputation among informed colleagues, Halley — an energetic astronomer and mathematician with a deep interest in celestial mechanics — traveled to Cambridge in August 1684 to ask Newton directly: what orbit results from an inverse-square force? Newton's reported immediate reply, "an ellipse", astonished Halley. Though Newton added that he would need to locate the calculation papers, his confident answer signaled that he already possessed the elusive mathematical derivation. Within weeks, Newton sent Halley a concise treatise, *On the Motion of Bodies in Orbit* (De Motu Corporum in Gyrum, circa 1684), giving a rigorous proof. Halley, instantly grasping its significance, resolved to see Newton's work developed and published in full.

That planets move in elliptical orbits was well-known to the scientific community, thanks to Johannes Kepler's laws published between 1609 and 1619. From meticulous analysis of astronomical observations, Kepler had deduced that planetary paths are ellipses with the Sun at one focus. The unresolved challenge was to identify and prove a physical force law that mathematically required such motion, while also accounting for Kepler's other laws. Newton's unique possession at that time was not the mere knowledge of elliptical orbits, nor even the hypothesis of an inverse-square attraction (which others had entertained), but the complete derivation — using his calculus methods — that an inverse-square central force necessarily produces conic-section orbits, specifically ellipses for bound planets.

Halley's reaction was one of immense excitement and determination. Realizing that this reclusive Cambridge professor held the key to a unifying explanation of celestial motion, Halley took on the role of champion. He persuaded Newton to prepare a comprehensive treatise, undertook the work of editing the manuscript, managed all dealings with the printer and the Royal Society, and, crucially, paid for the printing himself. The Royal Society's finances were drained — most recently by the expensive publication of Francis Willughby's illustrated *History of Fishes* (De Historia Piscium, 1686) — and without Halley's personal funding, Newton's book would have been delayed or imperiled. Halley may also have seen the advantage of bypassing possible institutional delays or interference, particularly from Hooke, ensuring that Newton's work appeared quickly and in full. Galvanized by Halley's unwavering support and the intellectual challenge, Newton set aside his other pursuits and devoted the next eighteen months to producing his masterpiece, *The Mathematical Principles of Natural Philosophy* (Philosophiae Naturalis Principia Mathematica, 1687). The publication not only revealed decades of Newton's private work, but also rekindled Hooke's priority claims regarding the inverse-square idea — prompting Newton to downplay his acknowledgements to Hooke in later editions.

Newton deliberately chose to write Principia primarily in a rigorous geometric style, building upon classical Euclidean methods. Geometry was still the most accepted language for fundamental proof, and Newton knew that presenting his results in this way was essential to persuade his contemporaries of their certainty. While the geometric form often masked the algebraic calculus he had used in discovery, it achieved its purpose. The impact of Principia was immediate and transformative: it established the three laws of motion and the law of universal gravitation, unified celestial and terrestrial mechanics under a single mathematical framework, mathematically derived Kepler's laws of planetary motion, explained the tides, the precession of the equinoxes, the orbits of comets, and more. It provided a coherent, predictive, mechanical model of the universe that became the bedrock of physics for over two centuries and stood as the ultimate exemplar of the mathematical-deductive scientific method.

One of the key conceptual tools underlying Principia was Newton's idea of *limits* — his "method of first and ultimate ratios". This was his solution to the logical problem of calculating instantaneous rates of change with vanishingly small increments without treating those increments paradoxically as both zero and non-zero. He sidestepped the philosophical tangles over infinitesimals by focusing on the definite, finite value that the ratio between these increments approached at the instant they 'appeared' or 'vanished'. While his method did not fully prove why these stable limiting values exist, it provided a convincing operational framework, keeping attention on the powerful physical results he could state and prove with certainty.

The publication of Principia transformed Newton from a relatively isolated Cambridge professor into an international intellectual superstar. He was elected Member of Parliament for Cambridge University (1689–90 and 1701–02), defending the university's rights against King James II. In 1696 he moved to London as Warden of the Royal Mint, becoming Master of the Mint in 1699. Newton applied his intellect to England's Great Recoinage, improved assaying methods, and pursued counterfeiters with zeal. His scientific authority was further cemented when he became President of the Royal Society in 1703, a post he held until his death in 1727. In this later period, he oversaw the second (1713) and third (1726) editions of Principia, published *Opticks* (1704, with two treatises on calculus appended), and continued his private studies in theology and chronology.

Initially, Leibniz, like virtually all leading scientists, recognized Principia as a work of epochal significance. He expressed admiration for Newton's achievement in establishing universal gravitation and mathematically deriving the system of the world, even publishing an influential (though anonymous) journal review praising its scope and power. Yet from early on he harbored deep philosophical objections, voiced over many years, to aspects of Newton's system. Both Leibniz and Huygens objected to Newton's reliance on *action-at-a-distance* — gravity acting without a contact mechanism — and to his concepts of absolute space and time. The book's rigorous but difficult geometric style also made its demonstrations challenging to follow. While Newton's physical results were studied intently, the style of presentation did not replace the algebraic calculus based on differentials and integrals that Leibniz had begun publishing only a few years earlier.

This situation created fertile ground for Leibniz's calculus to gain momentum, especially through the efforts of the Bernoulli brothers in Basel. They were not content to merely apply Leibniz's rules — they expanded them aggressively throughout the 1690s, extending the reach of their analysis to problems that surpassed earlier examples. They determined the shape of a hanging chain (the *catenary*), analyzed complex curves, advanced probability in connection with infinite series, and, most famously, tackled the *brachistochrone problem* — finding the curve of fastest descent between two points. The Bernoullis's work became a public demonstration of the breadth, power, and versatility of the Leibnizian calculus for all of Europe.

In June 1696, Johann Bernoulli posed the brachistochrone problem: to determine the curve between two given points (at different heights, not vertically aligned) down which an object would slide under gravity in the shortest time. The name itself came from the Greek *brachistos* (shortest) and *chronos* (time). The challenge quickly captured the attention of Europe's leading mathematicians. By 1697, several correct solutions were submitted, all identifying the curve as a cycloid, though using strikingly different methods.

Johann himself approached the challenge through a deeply clever analogy with the behavior of light. He knew of Pierre de Fermat's *principle of least time* and of Willebrord Snell's *law of refraction*, which states that when a ray passes between media of different speeds, the sines of its angles to the normal are in the same ratio as the speeds. This follows because sliding the crossing point a tiny amount along the boundary changes each leg's length at a rate equal to the sine of its angle to the normal, and the minimum-time path is where those length changes, weighted by the respective speeds, exactly cancel — yielding the ratio of sines to speeds. In everyday terms, as with a runner crossing from sand to pavement, the quickest route balances time in the slow and fast surfaces so that a small shift in the crossing point makes no difference to the total time.

Johann imagined the falling object as if it were a ray of light travelling through a stack of infinitesimally thin horizontal layers, each a *medium* in which gravity gave it a slightly greater speed than the one above. Just as a ray would refract at each interface according to the law of refraction, the object's path would bend continuously as its speed increased with depth, always adjusting to keep the total travel time minimal. Using Galileo's rule that the speed in free fall depends only on vertical distance, Johann expressed this continuous refraction as a mathematical condition — and from that found that the curve must be a cycloid.

Leibniz likely began from a similar physical insight to Johann's, drawing on the behavior of light. Unlike Johann, however, he applied the algorithmic machinery of his differential calculus directly, working with infinitesimals and derivatives to produce a *differential equation* — a relation connecting the slope of the sought curve at each point to the coordinates of that point — for the path the object must follow. His published solution gave a concise algebraic derivation — less immediately intuitive than Johann's analogy, but a direct demonstration of the calculus's power.

The problem reached Newton in England in January 1697, during his demanding new responsibilities at the Royal Mint. According to an oft-repeated anecdote recorded later by John Conduitt (Newton's assistant at the Mint and husband to his niece), Newton returned home exhausted, encountered the problem, and worked through the night to solve it before sleeping, sending his anonymous solution the next day. Though unsigned, its power and elegance were instantly recognizable to Johann, who reportedly declared "tanquam ex ungue leonem" ("I recognise the lion by his claw").

Suspicions about overlapping discoveries in calculus between Newton and Leibniz lingered beneath the surface, but had not yet erupted into open conflict. Newton's anonymous solution to a challenge from the Leibnizian circle may have sharpened the rivalry, showing that both "inventors" possessed closely related and formidable mathematical tools, and making the question of precedence still more pointed. The first public blow came soon after, around 1699, when Nicolas Fatio de Duillier (1664–1753), a Swiss mathematician closely associated with Newton in London, slipped a barb into a published tract — a suggestion that Leibniz had drawn his calculus from Newton's work, perhaps from material seen during earlier visits or in correspondence. The printed words crossed the Channel and reached Hanover, where Leibniz, incensed, sent a formal demand for retraction to the Royal Society. What had been a private undercurrent now broke the surface, drawing in allies and patrons on both sides and hardening into a public, nationalistically charged dispute that would divide the European mathematical world for decades.

Amid the growing turmoil, it was not Newton, nor Leibniz, nor Johann Bernoulli whose method would prove most consequential for the future of calculus. The most far-reaching approach came from Johann's elder brother, Jakob Bernoulli. Departing from the physical analogy to light and the direct differential calculation, Jakob treated the entire curve as the unknown, searching among all possible paths for the one of least time. Adapting Fermat's principle for finding maxima and minima, he argued that the correct path must be stable against any small change: if altering it slightly would shorten the time, it could not be the solution. This insight gave him an extra condition, beyond the problem's basic setup, that the curve would have to meet.

To carry out this plan, Jakob set up a way to treat the path like any other variable in a calculation for total travel time. At each horizontal position, he wrote the vertical coordinate of the unknown optimal curve as y, and then considered a slightly different curve, $y + \delta y$, where δy was a small variation used to test whether the time could be improved. This turned the total time into an integral — adding up the distance along the path divided by the speed at each point. Galileo's *law of free fall* gave the link between the curve's shape and the speed: the speed at any point depends only on how far the object has fallen vertically from the start. Applying Fermat's "flatness" condition in this more abstract setting, he required that, in the space of all possible paths, the graph showing how total time changes with the path would have a perfectly horizontal tangent — meaning no tiny change could make the time shorter. This became an equation that y had to satisfy — a differential equation relating the slope of the curve at each point to the point itself. Solving that equation gave the cycloid. Jakob's approach effectively launched what would become the *calculus of variations* — a distinct field built on Leibniz's calculus, extending its reach to an entire class of optimization problems involving unknown paths, or *functions*.

Building on this foundation, the Swiss mathematician Leonhard Euler (1707–1783), a student of Johann Bernoulli, emerged as the dominant mathematical figure of eighteenth-century Europe. His output was prodigious in both volume and influence — perhaps the most prolific and influential of any mathematician in history. He reshaped calculus and laid the groundwork for mathematical *analysis* — a rigorous approach to calculus built on precise definitions and logical proofs. His landmark textbooks included *Introduction to the Analysis of the Infinite* (Introductio in analysin infinitorum, 1748), *Foundations of Differential Calculus* (Institutiones calculi differentialis, 1755), and *Foundations of Integral Calculus* (Institutiones calculi integralis, 1768). These works became standard references for teaching while also extending mathematical theory far beyond the boundaries of the time.

Euler entered the University of Basel around age thirteen — not unusually early for the time — intending, like his father, to study theology. His fascination with mathematics quickly eclipsed all else, and he pursued it with relentless focus. He soon sought the guidance of Johann Bernoulli, then one of Europe's foremost mathematicians. Johann agreed to tutor him privately, almost certainly aided by a longstanding family connection: Euler's father, Paul, a Calvinist pastor, had been close friends with the Bernoullis since his own student years, when he attended Jakob Bernoulli's lectures while boarding in the Bernoulli household. In the late seventeenth century, it was common for theology and mathematics students in Basel to lodge with professors, receiving both lectures and personal instruction. Paul had boarded with Jakob; Johann, Jakob's younger brother, had himself once been a student lodger. Whatever the exact arrangement, Johann quickly recognized Euler's extraordinary talent and persuaded Paul to allow his son to pursue mathematics without constraint.

By 1726, Euler had completed his studies but found limited opportunities in Switzerland. The country's small university system offered few posts in mathematics, and even the Bernoulli family often competed internally for the scarce chairs in Basel. A far more ambitious prospect arose in Russia. Tsar Peter the Great's modernization program had established the Saint Petersburg Academy of Sciences to attract leading European scholars with generous salaries, prestigious appointments, and well-funded research facilities. Johann Bernoulli's sons, Nicolaus II (1695–1726) and Daniel (1700–1782), had already accepted positions there. When Nicolaus died suddenly of tuberculosis in 1726, Daniel and Johann moved swiftly to secure an invitation for Euler to join Daniel at the Academy — an opportunity that would launch his international career.

Euler's relationship with Daniel Bernoulli combined close collaboration with a healthy measure of competition. They exchanged ideas freely and worked together on difficult problems in mechanics, often spurring each other to new insights. One major shared interest was the emerging field of *hydrodynamics* — the mathematical study of fluid motion. Its importance was growing quickly, both in practical terms, for ship design, reducing fluid resistance, and building more efficient water wheels, pumps, and canals, and in theoretical terms, as mathematicians sought the general laws governing fluids. The tools of calculus were indispensable here, enabling precise descriptions of changing fluid velocities, pressures, and flow patterns, and making hydrodynamics a frontier area of research.

In 1733, Daniel left Saint Petersburg to accept a professorship in Basel, citing the political climate, censorship, and perhaps a sense of isolation. Euler, by contrast, adapted well to the Academy's environment. He found abundant opportunities, strong institutional support, and a high degree of academic freedom, allowing him to pursue research at an intense pace while also marrying and beginning a family. With Daniel's departure, the twenty-six-year-old Euler became the Academy's leading mathematician. His international reputation grew rapidly, fueled by a prolific stream of papers in the Academy's journal, the Commentarii, showcasing remarkable power and depth across many mathematical fields, setting the stage for the work that would soon bring him worldwide fame.

Meanwhile, in Britain, mathematics continued to evolve along a somewhat different course from the Continent, shaped by the lingering priority dispute and adherence to Newton's "fluxional" notation. Communication persisted across the Channel, but distinct emphases emerged. British mathematicians largely worked within Newton's framework, in which infinite series were central to handling many expressions. Without general symbolic rules for differentiating or integrating more complicated forms (such as roots or fractions), Newton's standard approach was to convert a formula into an infinite series — for example, by applying his *generalized binomial theorem* — and then apply his *fluxion* and *fluent* rules term-by-term. This reliance on series expansions as a first step fostered particular British skill in manipulating series, devising approximations from finite truncations, and exploring their properties.

Continuing in this tradition, Brook Taylor (1685–1731) published a general technique for expressing any variable quantity as an infinite series based on its derivatives at a given point in his fundamental *Direct and Indirect Methods of Incrementation* (Methodus Incrementorum Directa et Inversa, 1715). This *Taylor series* quickly became a cornerstone tool for analysis and approximation throughout Europe. James Stirling (1692–1770) addressed key problems in series summation and interpolation, and gave the useful *Stirling's approximation* for estimating large factorials. Using such techniques, Abraham de Moivre (1667–1754), a French Huguenot refugee in London and friend of Newton, published *The Doctrine of Chances* (1718), in which he identified the *normal distribution* — the *bell curve* — as an approximation to binomial distributions. His work also produced a formula linking complex numbers and trigonometry, anticipating Euler's imminent and more general unification of these ideas.

Steeped in the symbolic methods of Leibnizian calculus and equally at home with the infinite-series techniques that characterized the Newtonian school in Britain, Euler commanded a range of tools unmatched by any of his contemporaries. His arrival as Europe's new mathematical force was emphatically marked in 1735 with his celebrated solution to the *Basel problem* — to find the exact sum of the reciprocals of the squares ($1 + \frac{1}{4} + \frac{1}{9} + ...$), a challenge that had resisted the efforts of the previous generation, including Leibniz and Jakob Bernoulli. Euler's solution was characteristic of his style. Starting from the infinite-series expansion of $\frac{\sin(x)}{x}$ derived from its *Taylor series*, he compared the coefficients with those from an audacious infinite-product factorization based on its "zeroes". The result was the long-sought exact sum $\frac{\pi^2}{6}$.

In the process, he uncovered an unexpected link between infinite series and the distribution of prime numbers, helping to establish *analytic number theory*.

That triumph coincided with the publication of his two-volume *Mechanics, or the Science of Motion Presented Analytically* (Mechanica, sive Motus Scientia Analytice Exposita, 1736–1737), in which he reformulated Newtonian particle mechanics entirely in the notation and methods of Leibnizian calculus. Taken together, these achievements decisively established Euler's pre-eminence across Europe.

Over the following decades, Euler transformed calculus from a collection of intuitive procedures into a systematic discipline. His notation streamlined mathematical reasoning and allowed ideas to be communicated with unprecedented precision. He standardized the symbols for the base of natural logarithms (e), the imaginary unit ($i = \sqrt{-1}$), and the notation for functions ($f(x)$).

Euler defined a *function* as an expression connecting any number of variables and constants. By naming the formula itself in $f(x)$, he made the relationship a primary object of study, rather than only the resulting curve or quantities involved. This subtle but important shift allowed him to apply the same analytic methods to a wide variety of problems, unifying areas of mathematics that had previously been treated separately. Through the language he developed, Euler was able to precisely express a multitude of nuanced and sophisticated mathematical revelations.

His study of the number e (approximately 2.71828) was equally transformative. This constant, known as *Euler's number*, arises naturally in problems of continuous growth or decay, from population change to radioactive processes. Euler showed how it connects algebra with calculus through the exponential function e^x, whose rate of change is equal to itself — a defining principle in differential equations and mathematical modelling. Just as 0 is the unique number that leaves others unchanged by addition, and 1 is the unique number that leaves others unchanged by multiplication, e is the unique base whose exponential function e^x is unchanged by differentiation. Geometrically, the curve $y = e^x$ has the property that its slope at any point equals its height there, and that same value equals the area under the curve from zero to that point.

Differential equations describe how systems change by relating variable quantities to their own rates of change, with the aim of finding a function that exhibits the specified behavior. Any equation that connects the rate of change — or the rate of change of the rate of change, and so on — of a function to the function itself is a form of differential equation. Such equations — and their solutions — are inherently self-referential, or *recursive*. Consider a simple model of population growth: rate of population change = growth rate × current population. Because the rate of population change is directly proportional to the population itself, this is a differential equation. The exponential function is the simplest example of this kind of relationship: its rate of change is directly proportional to its current value, making it a natural building block in solving differential equations.

It was Euler who, in the mid-eighteenth century, organized the study of differential equations into a coherent branch of analysis. Early calculus provided tools for certain simple cases, but solving the diverse equations arising from physics and geometry still relied on problem-specific, ad hoc methods. Recognizing the need for general approaches, Euler began classifying these equations by their structure — for example, by "order", "linearity", "constant" versus "variable coefficients", and "ordinary" versus "partial". He then developed a versatile set of general solution methods, deriving foundational results such as the equations governing fluid flow (*Euler's equations*) and making major contributions to the theory of the *wave equation*. This framework, complete with clear classifications and broadly applicable techniques, presented in his comprehensive textbooks, firmly established differential equations as a central pillar of applied mathematical analysis.

Among Euler's most profound contributions was the elucidation of what is called *Euler's formula*: $e^{i\theta} = \cos\theta + i\sin\theta$. This identity revealed an unexpected unity between exponential functions, which describe growth and decay, and trigonometric functions, which describe waves and oscillations, through the bridge of imaginary numbers. It showed that these seemingly different types of functions are two aspects of the same relationship in the domain of *complex numbers* — numbers with both "real" and "imaginary" parts — providing a geometric interpretation for complex arithmetic and unifying algebra, trigonometry, and calculus. Beyond its theoretical beauty, the formula had immediate practical impact. It gave a direct way to interpret complex-valued exponential solutions of differential equations, translating them into the familiar sine and cosine forms used to model vibrations, wave motion, and other oscillatory phenomena.

In 1744, Euler produced the first systematic treatise on the *calculus of variations*, rigorously deriving the general equation and method for finding a function that optimizes an integral. He demonstrated its wide applicability by solving problems such as determining *geodesics* (shortest paths) on curved surfaces, finding minimal surfaces of revolution, and addressing *isoperimetric* problems (maximizing area for a fixed perimeter) and *isoepiphanic* problems (minimizing perimeter for a fixed area). With this technique, Euler showed that the shortest path on a sphere is an arc of a great circle; that a soap film spanning two identical rings forms a *catenoid*, the surface generated by revolving a hanging-chain catenary; and that the circle uniquely maximizes enclosed area for a given perimeter, and conversely minimizes perimeter for a given area. In each case, he wrote the quantity to be optimized as an integral, then applied the flatness condition at a maximum or minimum — setting the derivative of the integral to zero and solving the resulting differential equation for the function that satisfies it.

In one of his investigations, Euler analyzed the bending of elastic beams under compression by framing the deformation as a "variational" problem. He formulated it on the principle that the beam would assume the shape that minimizes its stored elastic potential energy, expressed as an integral of the square of the beam's *curvature* — the rate of change of its angle — along its length. The physical intuition was that each infinitesimal slice of the beam tries to restore itself to straightness in proportion to its curvature at that point. Using the calculus of variations, he systematically derived differential equations describing the curvature under applied loads. The solutions to these equations — the actual bent shapes — could not, in general, be expressed in familiar elementary functions such as polynomials, trigonometric functions, or logarithms. Instead, they involved *elliptic integrals*, a class of functions that had first appeared in problems such as finding the arc length of an ellipse, and which Euler himself had extensively studied and advanced. His analysis allowed him to classify the wide variety of possible equilibrium shapes, or *elastica*, from simple bends to complex undulating and even self-intersecting curves. He also determined the critical load beyond which a straight column becomes unstable and buckles into one of these shapes, deriving the *Euler buckling formula*, still fundamental in structural engineering.

Across his career, Euler also pioneered what are called *optimal control problems*. He derived time- and distance-minimizing paths for particles and pendula, including the brachistochrone, and recast Fermat's principle of least time in optics within his variational framework. He formulated the equilibrium shapes of fluid surfaces and extended his methods to continuous media, introducing velocity and pressure fields to derive the *Euler equations* for fluid flow. Applying the same tools to ballistics, he optimized projectile trajectories under gravity and air resistance, producing explicit range formulae. In ship design, he posed the hull-shaping problem as a drag-minimization integral and derived the differential conditions an optimal hull must satisfy, laying the groundwork for modern hydrodynamic optimization.

Euler carried these methods into celestial mechanics, writing *action* integrals for orbital motion, analyzing special two-body solutions, and addressing a restricted three-body lunar problem. Using *perturbation series*, he refined Earth–Moon–Sun predictions to a level of accuracy unmatched at the time. Through these wide-ranging achievements, Euler wove a unifying analytical thread that linked mechanics, optics, fluid dynamics, engineering, and astronomy, transforming both the calculus of variations and *perturbation theory* into universal tools for understanding the natural world.

Before Euler, Newtonian mechanics concentrated largely on the motion of point masses or highly simplified bodies. Extending this framework to the full three-dimensional motion of a rigid body was a major challenge. Euler provided the first comprehensive mathematical treatment of such motion — for objects that, unlike fluids, maintain a fixed shape and internal structure as they move. He showed that their motion could be analyzed as the combination of translation (movement of the center of mass, like a point particle) and rotation about that center.

Euler's key insight into rotation came from the body's rigidity: because its shape does not change, its distribution of mass relative to its own axes — described by its *moments of inertia* — remains constant. This fixed internal structure leads to conserved quantities when no external twisting forces (torques) act: the body's total angular momentum (its quantity of rotation) remains constant in direction and magnitude in space, and its rotational kinetic energy also remains constant. Euler derived the fundamental differential equations — called *Euler's equations of rigid body motion* — showing that to conserve angular momentum in space, the body's angular velocity (its spin rate and direction relative to itself) must generally change in complex ways. These equations explain phenomena such as the wobbling (*precession*) and nodding (*nutation*) of spinning objects. He also introduced *Euler angles* as a systematic way to describe a body's changing orientation. The contrast with fluids is sharp: fluids can deform internally, redistributing momentum and energy in ways that rigid bodies cannot, and are described by entirely different equations.

In optics, Euler was a prominent eighteenth-century advocate of the wave theory of light, opposing Newton's prevailing "corpuscular" view. Building on Christiaan Huygens's ideas, he argued that waves provided a better explanation for phenomena such as diffraction, and proposed that different colors correspond to different vibration frequencies in a hypothetical *luminiferous ether*. Beyond theory, he addressed the practical problem of chromatic aberration in refracting telescopes — the tendency of lenses to focus different colors at different points. Newton had been pessimistic about correcting this, but Euler, applying the laws of refraction and dispersion, proved mathematically that it was possible. He showed that by combining lenses made of different types of glass, such as "crown" and "flint", with differing refractive and dispersive properties, one could bring multiple colors to a common focus. This analysis established the design principles behind achromatic lenses, a major advance in optical instrumentation.

Beyond essentially defining modern *analysis*, Euler laid foundational work in graph theory, number theory, and topology. He was far from an isolated figure: he corresponded and collaborated with nearly every leading mathematician of his era, influencing the next generation and fostering a vibrant environment for mathematical progress.

Among his contemporaries was Alexis Clairaut (1713–1765), a prominent French mathematician and physicist who often worked on the same problems as Euler, especially in celestial mechanics. Both made major, and sometimes competing, advances in *lunar theory* — predicting the Moon's intricate orbit — and in determining the Earth's shape under the combined effects of gravity and rotation. This overlap led to direct competition, frequently intensified by the prize contests of the Paris Academy of Sciences, which both men won multiple times for related work.

When Euler and others encountered difficulty in verifying Newton's law of gravitation for the three-body dynamics of lunar motion, Clairaut recognized that the problem lay not in Newton's law but in the sensitivity of the calculations to small errors. His remedy was to introduce higher-order approximations, an approach that preserved Newton's framework. Despite their rivalry, the two men maintained professional correspondence and mutual respect, each acknowledging the other's analytical prowess.

Amid this vibrant and often competitive climate, the twenty-six-year-old Parisian polymath Jean le Rond d'Alembert (1717–1783) — largely self-taught — published his first major work, *Treatise on Dynamics* (Traité de dynamique, 1743). In it, he proposed a principle for analyzing systems in motion, introducing the concept of *dynamic equilibrium*, which Euler at once praised as "ingenious and very useful". D'Alembert's approach built on the existing principle of *virtual work*, a method for analyzing systems in *static equilibrium* (at rest). This principle considers the *work* done — the product of a force and the directed distance over which it is applied. For a static system, it states that if one imagines giving the system an instantaneous, infinitesimal (virtual) nudge consistent with its constraints (its allowed motions), the total virtual work is zero. This arises because the ideal constraint forces — such as the normal reaction of a surface or the tension in a rigid rod — act perpendicular to these allowed nudges and therefore do no work in their direction of motion.

D'Alembert's principle was a major advance in the mathematical modeling of motion. Before it, directly applying Newton's laws to constrained systems was often prohibitively difficult, forcing physicists to simplify by relying on additional equations derived from identifiable conserved quantities — typically total vector *momentum*, central to Newtonian mechanics, or the scalar *vis viva* ("living force", roughly *kinetic energy*), central to Leibnizian thought. D'Alembert's method offered a way to frame dynamics without first choosing sides in the long-running Newton–Leibniz debate over which of these quantities was the more fundamental "conserved quantity of motion". By grounding his formulation in an instantaneous balance of work, he provided a systematic framework for tackling a wider class of problems, independent of earlier philosophical divisions.

Around 1746, d'Alembert introduced the *one-dimensional wave equation* describing the motion of a vibrating string, and gave its general solution in the elegant form of a sum of pulses traveling in opposite directions. This was a landmark result, effectively launching the mathematical study of wave propagation and producing a *partial differential equation* — an equation that relates derivatives of a solution function with respect to multiple "input" variables, thus equating rates of change in one input to rates of change in another.

D'Alembert derived the wave equation by modeling a string under constant tension and examining how a small disturbance is restored toward equilibrium. He divided the string into infinitesimal slices, balanced the forces on each slice, and, assuming the displacements were small, simplified the geometry so that only first-order effects in the slope were kept. He found that any proposed solution must have the property that its second derivative with respect to time (its *acceleration*) is proportional to its second derivative with respect to position (its *curvature*).

He then showed that if one adds together any two "single-input" functions — one sampled with its input shifting to the left over time and the other sampled with its input shifting to the right — the result will always satisfy this condition. In the formal language of mathematics, the wave equation can be expressed concisely as $\frac{\partial^2 u}{\partial t^2} = c^2 \frac{\partial^2 u}{\partial x^2}$, which has any function of the form $u(x,t) = f(x-ct) + g(x+ct)$ as its solution, where f and g represent arbitrary functions.

Euler and Daniel Bernoulli responded to d'Alembert's vibrating-string memoir almost immediately, and their differing reactions sparked a debate over what mathematicians should be willing to call a function. Writing from the Berlin Academy in 1749, Euler accepted d'Alembert's partial differential equation but extended it far beyond a single string: he derived analogous formulas for stretched membranes and for sound waves in air, and solved them by separating variables into trigonometric "modes". Bernoulli, drawing on the acoustic fact that a plucked string produces a "fundamental" tone plus higher *harmonics*, argued that any initial shape can be expressed as an infinite sum of standing sine waves. Joseph Sauveur (1653–1716) had demonstrated this principle experimentally in the early eighteenth century by sprinkling fine powder on a horizontally vibrating string under tension: the powder is shaken from the moving regions and collects at the fixed "nodes" located at fractional lengths of $\frac{1}{2}, \frac{1}{3}, \frac{1}{4}$, and so on.

The three men quickly clashed over the scope of this representation. Euler welcomed Bernoulli's sine modes as particular solutions but warned that the terms of a series can be differentiated only when the curve they define comes from a *convergent* expression — one that does not grow without bound. Bernoulli insisted that even sharply peaked plucks could be represented by his series. D'Alembert, by contrast, rejected such "cornered" shapes outright, arguing that the *wave equation* involves second derivatives, which cannot be taken at sharp corners. Between 1749 and 1755 they traded a flurry of examples and counterexamples, gradually refining the meaning of "arbitrary" and "smooth" in mathematical analysis.

The controversy pushed the development of new tools. To support his smoothness requirement, d'Alembert wrote the Encyclopédie entry *Limit* (Limite, 1754), defining a *limit* as a quantity that a variable approaches "as close as one chooses". In 1768 he introduced the *ratio test* for series convergence: if the long-run ratio of successive terms is less than one, the sum converges to a limit. Euler, less formal but equally influential, showed that Bernoulli's sinusoidal series reproduces d'Alembert's travelling-pulse solution whenever the functions are twice differentiable, and suggested approximating any rough profile by a smooth one to any desired accuracy — an intuitive, limit-like argument. These exchanges laid the groundwork for the rigorous theory of functions that Jean-Baptiste Joseph Fourier (1768–1830), Augustin-Louis Cauchy (1789–1857), Johann Peter Gustav Lejeune Dirichlet (1805–1859), and Bernhard Riemann (1826–1866) completed in the next century.

Initially working in Turin, Italy, Joseph-Louis Lagrange (1736–1813) initiated contact with the famous Euler in 1755 by sending him early work containing an elegant formulation of the calculus of variations using "δ-notation". Euler immediately recognized Lagrange's genius, enthusiastically praised and promoted his work, and incorporated Lagrange's ideas — with full credit — into his own later writings on the subject. He even withheld some of his own related results to allow Lagrange to publish first. Over years of extensive correspondence, Lagrange deeply impressed Euler, eventually succeeding him as director of mathematics at the Royal Prussian Academy of Sciences in Berlin.

Lagrange's *Analytical Mechanics* (Mécanique Analytique, 1788) represented the culmination of decades of work and an ambitious vision. His aim was to reconstruct the entirety of classical mechanics — both statics and dynamics — from a single fundamental principle, relying purely on mathematical analysis and famously boasting in the preface that "no diagrams will be found in this work". This drive stemmed from a desire for elegance, logical unity, and generality, seeking to transform the science of motion into operations as systematic as solving algebraic equations.

Central to Lagrange's approach was the quantity he denoted $L = T - V$ — later called the *Lagrangian* — defined as the kinetic energy T minus the potential energy V of a system. He demonstrated that, expressed in terms of L, the complete dynamics of a physical system could be recovered from a single principle. His analysis depended upon d'Alembert's principle, which had required that any imagined "virtual" nudge of a system must fit the assumed constraints (for example, that links in a chain can move but must keep the same length). The challenge was that, when carrying out the calculations, those restrictions would not necessarily remain satisfied unless they were built into the mathematics. Lagrange devised two general strategies to make the restrictions part of the equations from the start, ensuring that any hypothetical nudge automatically respected them.

The first strategy was his systematic development of the *method of undetermined multipliers*, building upon earlier ideas from Euler. A convenient set of (possibly interdependent) coordinates was chosen to describe the system, and the constraints were stated explicitly as algebraic equations. For each constraint, a *multiplier* — an additional unknown variable — was introduced. This enlarged the system of equations: the original equations of motion, now with added terms involving these multipliers, were solved together with the constraint equations. This approach not only produced the motion consistent with all restrictions, but also revealed the multipliers themselves as measures of the forces of constraint.

The second strategy showed that, by intentionally choosing coordinates aligned with the system's true degrees of freedom, the computational effort could be greatly reduced. Lagrange called any such set of coordinates *generalized coordinates*. Using them, he rewrote the inertial force in algebraic form as a difference of derivatives of the kinetic energy function. Building again on the work of Euler and Daniel Bernoulli, he separated applied forces into "conservative" and "non-conservative" components. A conservative force does the same work regardless of the path taken between two points, and Lagrange expressed it as the rate of change of potential energy with respect to (generalized) position. His decisive step was to translate d'Alembert's principle into a differential equation: in any set of generalized coordinates, the time rate of change of the rate of change of L with respect to velocity equals the rate of change of L with respect to position, plus any non-conservative forces. The remainder of his text shows how this single equation — applied to the appropriate form of L — could solve the entire range of classical mechanical problems, from simple machines and planetary orbits to vibration theory and ideal-fluid flow, all without a single geometric diagram.

Like Euler, Lagrange's work extended far beyond a single topic, reshaping multiple branches of mathematics. In algebra, his study of polynomial equations — algebraic expressions with variables raised to various powers — revealed that the symmetries among their possible solutions held the key to knowing whether a general formula for the solutions could exist. This insight laid foundations for *group theory*, the abstract study of symmetry itself. His contributions to number theory were equally significant: he provided the first complete proof of Fermat's claim that every positive integer can be expressed as the sum of at most four squares. Lagrange's proof used algebraic identities — notably one due to Euler connecting sums of four squares — together with a refined *method of descent* reminiscent of techniques the ancient Greeks had used to prove results such as the irrationality of $\sqrt{2}$. He also brought his analytical skill to probability theory, helping to secure its mathematical foundations through calculus-based methods for analyzing chance and data. In celestial mechanics, he developed sophisticated tools to calculate the gravitational interactions and long-term orbital stability of planets and moons, and famously identified the gravitationally stable *Lagrangian points* in orbital systems.

By the 1780s, Paris — despite France's deepening fiscal crisis and growing social unrest — had retained its standing as Europe's preeminent scientific capital. Other cities, such as Berlin, had enjoyed periods of brilliance, often centered around a figure like Euler (until 1766) and then Lagrange, due to specific royal patronage such as that of Frederick the Great. Paris, however, offered a unique concentration of talent within its Royal Academy of Sciences and a vibrant, critical intellectual culture. It was home to luminaries shaping the course of science: Antoine Lavoisier, for example, was in the midst of his chemical revolution, bringing quantitative rigor to chemistry. The allure of Paris was immense for any leading mind.

When Frederick the Great died in 1786, altering Lagrange's circumstances in Berlin, Louis XVI extended a prestigious invitation for him to join the Academy in Paris in 1787, with lodgings in the Louvre — then a residence for many artists and scholars under royal favor. For Lagrange, this was a highly attractive proposition that would mean returning to a French cultural milieu (he had French ancestry) at what was widely perceived as the center of the scientific world, likely underestimating the imminent revolutionary storm, even if the air was thick with calls for reform.

By this time, Pierre-Simon Laplace (1749–1827) was already a commanding figure within the Parisian scientific world. Arriving in Paris around 1768 as a young man from Normandy with a letter of introduction to Jean le Rond d'Alembert, Laplace quickly impressed d'Alembert with his exceptional mathematical talent, reportedly solving a set of difficult "audition pieces" overnight. D'Alembert's support was decisive: he secured Laplace a professorship at the Military Academy (École Militaire), providing him both a stable base to launch his research and after-hours access to the Paris Observatory — access that would prove essential for his next feats.

Laplace set his sights on some of the most profound and vexing problems in astronomy: the apparent *secular acceleration of the Moon* — the observation that the Moon's orbital motion appeared to be gradually speeding up over centuries, raising fears it might eventually spiral away or crash into the Earth — and the long-term aberrations in the motions of Jupiter and Saturn. These giant planets were not following their predicted Keplerian orbits precisely: Jupiter appeared to speed up over centuries, while Saturn seemed to slow down. The possibility that these deviations could destabilize the Solar System had even led Newton to speculate about occasional divine intervention to preserve cosmic order.

By 1773, Laplace had mastered the "variation of parameters" method pioneered by Lagrange. Starting from a planet's dominant motion as a Keplerian ellipse about the Sun, he allowed the elements of that ellipse to vary under the gravitational pulls of other planets. The resulting *perturbation functions* for those orbital elements could be expressed as explicit differential equations. Solving them showed that the motions must ultimately be periodic, not runaway. This work earned the twenty-four-year-old Laplace a seat in the Academy of Sciences.

Over the next decade, he assembled the tools to prove his theory empirically. He developed a fledgling probabilistic framework to evaluate the reliability of telescope readings, secured an agreement with the Paris Observatory's director for first access to the latest positional data, and obtained small Academy stipends to pay a handful of junior "computers" to expand his vast tables of numerical coefficients. Each computed term was tagged with two measures — its maximum possible size and the slowness of its variation. Only corrections both large enough to matter and slow enough to pile up over centuries were carried forward; all others were discarded.

Using this careful filtering, Laplace completed the formidable calculations and, in 1786, demonstrated that Jupiter and Saturn do not drift apart indefinitely but instead oscillate by less than a degree in a graceful nine-hundred-year cycle. This Great Inequality was explained entirely within standard Newtonian mechanics, carried out to an unprecedented level of precision.

While the clerk team was busy crunching endless coefficient tables, Laplace saw that mapping the celestial dance of planets and moons with ever-greater precision would require more powerful mathematical tools. He focused on the potential energy term V as the central, unifying quantity for understanding gravity. His crucial step was to show how this single, space-filling *field* could offer a complete picture of a celestial body's gravitational influence.

Instead of wrestling with countless force *vectors*, he worked with this more manageable *scalar* — a single number at each point in space — whose *gradient* (its rate of change with respect to position) revealed both the direction and the strength of the gravitational force there. This conceptual shift promised a cleaner and more powerful way to attack the Solar System's complexities.

Building on this, Laplace formulated a fundamental rule — a differential equation called *Laplace's equation* ($\frac{\partial^2 V}{\partial x^2} + \frac{\partial^2 V}{\partial y^2} + \frac{\partial^2 V}{\partial z^2} = 0$) — which the potential V must satisfy everywhere in empty space. It says that a gravitational potential in empty space has no net curvature: the field is perfectly "flat" where there is no mass present to curve it. Mass generates the curvature of the potential field; space carries the potential evenly. Laplace's equation, developed in a series of memoirs between 1782 and 1785, became the cornerstone of *potential theory*.

To apply it to real celestial bodies, which are not perfect spheres but often bulge at their equators (like Earth or Jupiter), a way was needed to represent more complicated gravitational fields. At the same time, Adrien-Marie Legendre (1752–1833) was developing special functions — called *Legendre polynomials* — that were perfect for describing such flattened, axially symmetric shapes. Just as Bernoulli had shown that the complex vibration of a string could be built from simple sine waves — its fundamental *harmonics* or *modes* — Laplace showed that the gravitational potential outside any planet could be written as an infinite series of more general *spherical harmonics*. These are the natural, fundamental patterns, or "modes", that solve Laplace's equation on a sphere. Each harmonic adds a distinct shape to the overall field, and the series coefficients give the "strength" or "amplitude" of each mode, allowing a detailed fingerprint of a planet's gravitational field.

The gravitational potential can be pictured as a smooth, undulating landscape, where the "height" at each point represents the value of the potential. The gravitational force is like a ball rolling across this landscape — it always rolls downhill along the "steepest" slope (the *gradient*). For the steepness to be well-defined everywhere, the potential landscape must be smooth; it cannot have infinitely sharp cliffs or jagged points where the slope is undefined. This is why the potential V must be differentiable. Laplace's equation encodes a profound rule about this landscape in empty space: at any point, the potential equals the average of the values immediately surrounding it. This "averaging rule" is the essence of why the field is smooth. In regions without mass — the sources of gravity — the potential landscape will not spontaneously form peaks or valleys on its own; such features can only be "propped up" or "carved out" by nearby matter. To Laplace, this intuitively meant that gravitational influence spreads itself as evenly and placidly as possible when far from its sources.

This "averaging" nature dictated by Laplace's equation also hints at why the potential field in a given region, once its values are fixed at the boundaries — for example, by distant stars or the surfaces of nearby planets — should settle into one unique configuration, much like a stretched soap film attached to a wire loop finds a single minimal-energy shape. While formal uniqueness proofs would come later, this inherent predictability was vital for Laplace's calculations. A field that meets this "value-equals-average" rule is called *harmonic*. Harmonic functions appear whenever a quantity diffuses, relaxes, or stretches out until every local excess has been smoothed away. Gravity outside mass is only one — and historically the first — famous member of that broader family.

The French Revolution, erupting in 1789, threw the established scientific order into a period of both peril and radical reorganization. Institutions like the Royal Academy of Sciences were suppressed in 1793, and the Reign of Terror brought with it genuine danger for intellectuals. Antoine Lavoisier's fate demonstrates this danger: despite his immense scientific contributions and his work for various reform committees (including early involvement in the metric system), his past role as a private tax collector under the old monarchy sealed his doom. He was guillotined in May 1794, with the infamous — though possibly apocryphal — remark attributed to the judge: "the Republic has no need of scientists". Joseph-Louis Lagrange, who had collaborated with and deeply respected Lavoisier, is said to have lamented, "it took them only an instant to cut off that head, and a hundred years may not produce another like it". Yet the Revolution also carried a fervent belief in reason and utility, leading to the creation of new institutions such as the École Polytechnique and the Institut de France, where science was to serve the nation.

Laplace, for his part, managed to continue his work and even enhance his influence through these tumultuous decades. He actively participated in revolutionary reforms, most notably serving on the commission that established the metric system — a hallmark of revolutionary rationality. While his contemporary Adrien-Marie Legendre faced greater personal and financial hardship during this period, focusing more narrowly on his research, Laplace's pragmatism saw him appointed as an examiner and professor in the new educational system.

His standing rose even higher under Napoleon Bonaparte. Years earlier, as an examiner for the Royal Artillery Corps, Laplace had conducted the final examination for the sixteen-year-old cadet Napoleon at the École Militaire in Paris. Napoleon passed, qualifying as a second lieutenant. Years later, as First Consul and then Emperor, Napoleon clearly remembered and respected Laplace's intellect. He appointed Laplace Minister of the Interior in late 1799, though the political career was famously short-lived. Napoleon dismissed him after just six weeks, reportedly quipping that Laplace "sought subtleties everywhere, had only problematic ideas, and, in short, carried the spirit of the 'infinitesimals' into administration".

Despite this brief and unsuccessful foray into high office, Napoleon continued to hold Laplace in high esteem, appointing him to the Senate and later bestowing on him the title of Count of the Empire. This patronage supported Laplace's publication of the remainder of his masterwork, *A treatise of celestial mechanics* (Traité de mécanique céleste, 1798–1825). This five-volume work was the culmination of Laplace's efforts, monumentally synthesizing all of mechanics to explain the intricate motions and argue for the stability of the Solar System with unprecedented precision, while at the same time profoundly advancing mathematical techniques and the solution of partial differential equations.

Another key French mathematician of this era was Gaspard Monge (1746–1818), a fervent revolutionary whose major intellectual contribution was *descriptive geometry*. Descriptive geometry provided a rigorous system of projections — typically top and front views — that allowed engineers and architects to represent three-dimensional objects accurately on a two-dimensional plane, solve spatial problems graphically, and communicate complex designs with clarity. It proved invaluable for industry, engineering, and architecture, making it possible to design and construct complex structures such as fortifications, machines, and ship hulls using only two-dimensional drawings.

A committed supporter of the Revolution, Monge devoted his energy to serving the new republic. He was instrumental in creating the metric system and even briefly served as Minister of the Marine. Equally important, he played a pivotal role in reforming scientific education, serving as a principal architect and professor at both the short-lived Normal School (École Normale, founded in 1795 to train teachers for the republic) and, most importantly, the Polytechnic School (École Polytechnique, founded in 1794).

From his position of authority, Monge ensured that invaluable intellects such as Lagrange were not only protected during the Revolution but placed at the heart of these new institutions. Lagrange became the first professor of analysis at the École Polytechnique, where his lectures and rigorous approach profoundly shaped the school's mathematical foundations. Monge himself taught there, fusing infinitesimal calculus with three-dimensional descriptive geometry and showing how to express surfaces, *normals*, and curvatures algebraically, then "read" the physics from their differential relations. Laplace also contributed to teaching and curriculum design, while Legendre took a quieter role serving as a permanent mathematics examiner.

The École Polytechnique quickly became a breeding ground for a new wave of brilliant mathematicians. Augustin-Louis Cauchy, perhaps the most emblematic figure of this French school, redefined analysis by introducing rigorous concepts of limits, continuity, and convergence, and by systematically developing *complex function theory*. His contemporaries and fellow alumni carried this precision and ambition into many branches of mathematics. Louis Navier (1785–1836), an engineer, applied advanced analysis to fluid motion, co-developing the *Navier–Stokes equations*. Jean-Victor Poncelet (1788–1867) revitalized projective geometry, while Jean-Baptiste Joseph Fourier advanced powerful *series methods* for the study of heat, broadening the mathematical toolkit available to physicists and engineers.

Siméon-Denis Poisson (1781–1840) was another one of the school's outstanding products. He built directly on Laplace's vision of using potentials as a unifying way to describe forces in nature, finding new ways to link the behavior of a field in space to the conditions imposed on it at the boundaries. This approach allowed him to frame problems in heat, gravity, and electrostatics within a single, coherent mathematical structure. He also recast Lagrange's mechanics in a clearer, more general form that could handle complex systems and stresses, work that would resonate deeply in later developments of continuum mechanics and mathematical physics.

Together, these mathematicians thrived in a structured, state-sponsored environment that was unmatched anywhere else in Europe. The French system's deliberate cultivation of mathematical talent and its close integration with engineering, science, and national institutions made Paris the gravitational center of much of the continent's mathematical life.

Beyond France, new channels for the exchange of ideas were multiplying. Scientific journals were growing in number and reach, carrying discoveries and methods swiftly across borders and complementing the older role of national academies. At the same time, the idea of the university itself was being reshaped. In the German states, a different, equally ambitious model was taking root — one that emphasized the unity of teaching and research, academic freedom, and the pursuit of knowledge for its own sake. Most famously embodied in the University of Berlin (Universität zu Berlin, founded 1810), this Humboldtian model fostered deep theoretical inquiry and encouraged the building of entire disciplines from first principles. In this fertile academic atmosphere, new approaches to mathematics and the physical sciences would take shape, setting the stage for major advances in fields such as magnetism, electricity, and the unification of physical laws.

A Most Natural Symmetry

In late eighteenth century Europe, physics was still spoken in the language of Newtonian mechanics. Gravitation, acting invisibly and instantaneously across empty space, was the gold standard for explanation; celestial mechanics stood as its showcase. When Benjamin Franklin, Charles-Augustin Coulomb, and others began to measure electrical forces, the natural step was to describe them with the same inverse-square law that had worked so well for planets. Charges played the role of masses, and attraction or repulsion seemed nothing more than gravity in a different costume.

The resemblance did not last for long. Electric sparks leapt, currents surged, and magnets snapped to attention — behavior far livelier than the steady march of planets. Luigi Galvani's twitching frog muscles hinted that electricity coursed through living tissue; Alessandro Volta's chemical "pile" of 1800 produced a steady stream of current that experimenters could use at will. Coulomb's torsion balance confirmed that both electric charges and magnetic poles obey an inverse-square rule, yet new puzzles appeared: opposite charges could neutralize one another, currents seemed able to create magnetism, and induced currents arose without direct contact. The tidy gravitational picture began to fray.

The mathematical response, however, continued to follow Newton's pattern. Pierre-Simon Laplace and Siméon-Denis Poisson expressed these forces in terms of a single abstract quantity: the potential. A potential was a way of describing how the influence of charges or poles spread through surrounding space, with the actual force obtainable by calculating how steeply that potential changed. This approach captured static arrangements of charge or magnetism with impressive accuracy, but it assumed that changes occurred everywhere at once. Nothing in the mathematics accounted for the sudden kick of induction or for the storage and transfer of energy that experiments were beginning to reveal. Electricity and magnetism remained side by side like estranged relatives — plainly connected, yet still without a common framework.

Carl Friedrich Gauss entered this scene from a different angle. Where Laplace and Poisson had refined Newton's style of reasoning into differential equations for static potentials, Gauss showed that the same framework could be grasped in geometric terms: the total effect of a distribution of charges or poles could be measured by the *flux* through an enclosing surface. This was more than a change of language. It turned potential theory from an abstract calculus into a principle that tied sources, geometry, and measurement together with new clarity.

Yet Gauss's insight, sharp as it was, remained a still picture. It described how forces balanced in space, but said nothing about how magnetic influence spread or how electrical energy seemed to vanish in one place and reappear in another. Those questions grew urgent in the early 1830s, with Faraday's discovery of induction and with the precision instruments then multiplying in European laboratories. From that decade onward, a new framework began to take shape — one that could treat fields not merely as static patterns but as evolving actors, capable of carrying energy and momentum in their own right. Piece by piece, the elements came into view: boundary methods that extended Gauss's surface idea, energy arguments that demanded local accounting of power, and lines of force that gave a tangible alternative to abstract action-at-a-distance.

Across the Rhine from France, a distinct educational experiment was taking shape. Where the post-Revolutionary French "écoles" trained engineers in fixed courses designed for state service, the new University of Berlin, founded in 1810 under Wilhelm von Humboldt, treated scholarship itself as a national project. Its guiding principles were wide-ranging study (*wissenschaft*), freedom to teach and to learn (*lehrfreiheit* and *lernfreiheit*), and the inseparability of research from instruction. Seminars and laboratories supplemented the lecture, and the requirement that each candidate contribute an original dissertation gave rise to the modern research doctorate. This Humboldtian model furnished German science with a steady supply of investigator-teachers, just as France supplied engineer-administrators, establishing complementary — and at times rival — paths toward the physics of the nineteenth century.

Within this evolving landscape, Gauss emerged as a mathematical prodigy whose early development owed more to private study and patronage than to institutional reform. Yet in his later career at Göttingen, Gauss came to embody the strengths of the German style: a commitment to first principles, an insistence on precision, and an ability to weave pure mathematics and physical measurement into a seamless whole. His reputation spread across fields — from number theory and astronomy to geodesy, magnetism, and differential geometry — setting a benchmark for rigor that others would emulate.

Born in Brunswick to a working-class family, Gauss's precocious abilities became legendary. He was said to have corrected his father's payroll as a child and to have astonished teachers by summing an arithmetic series with extraordinary speed. With the patronage of the Duke of Brunswick, he pursued an education that quickly surpassed the curriculum, devouring the works of Newton, Euler, and Lagrange while subjecting them to his own critique. By his late teens he had already produced discoveries that would reshape mathematics: the constructibility of the regular seventeen-sided polygon, the foundations of modular arithmetic, the conjecture of the prime number theorem, and the independent formulation of least squares. This combination of self-reliance, institutional support, and breathtaking originality made him the prototype of the nineteenth-century German scholar.

The publication of his early masterpiece in number theory, *Arithmetical Investigations* (Disquisitiones Arithmeticae, 1801), when he was twenty-four, transformed the field from a collection of scattered results into a systematic discipline. In it Gauss introduced the arithmetic of remainders (*congruences*) as a unifying tool, gave the first complete proof of the *law of quadratic reciprocity* — a relation among prime numbers that had resisted Euler and Legendre — developed a general theory of *binary quadratic forms* (expressions of the form $ax^2 + bxy + cy^2$), and extended the theory of *cyclotomy* (*dividing the circle*), which included his earlier construction of the regular seventeen-sided polygon. The book's austere, compressed style sometimes obscured the scale of its innovations, but it established Gauss as one of Europe's leading mathematical minds.

That same year brought him into the practical sciences. The newly discovered asteroid Ceres had vanished behind the Sun after only a few weeks of observation. Drawing on sparse data, Gauss used a new method of least squares, along with refined orbital calculations, to predict its reappearance with striking accuracy. The success brought him recognition beyond mathematics and, in 1807, an appointment at the University of Göttingen as Professor of Astronomy and Director of its observatory — posts he would hold for life.

Astronomical work soon led him to geodesy, directing a long survey of the Kingdom of Hanover. The demands of measuring across great distances prompted him to design special instruments, such as the *heliotrope*, which used reflected sunlight as a surveying signal. They also led him deeper into the geometry of surfaces. His *General Investigations of Curved Surfaces* (Disquisitiones Generales circa Superficies Curvas, 1827) introduced the concept of intrinsic curvature, showing in his *Theorema Egregium* (remarkable theorem) that a surface's curvature is an internal property, unchanged when the surface is bent without stretching. This discovery opened the way to non-Euclidean geometry and, more than a century later, provided essential ground for the theory of general relativity.

In September 1828, Alexander von Humboldt (1769–1859) gathered Europe's leading natural philosophers in Berlin. Celebrated explorer, confidant of kings, and tireless promoter of science, he had just returned from a global tour that carried him from the Andes to the Urals. He used the congress to argue that Earth's magnetic field deserved the same systematic attention astronomers gave the heavens. Compass charts, he warned, were drifting into error; sudden "magnetic storms" could swing a ship's needle by whole degrees; and only simultaneous observations around the globe could reveal the cause. To this cause he sought to enlist Carl Friedrich Gauss. Normally reluctant to leave Göttingen, Gauss made a rare exception, persuaded by Humboldt's personal invitation to stay as his guest, and by the intellectual ferment promised.

Among the young men presenting was Wilhelm Weber (1804–1891), who displayed finely crafted apparatus of his own design for measuring magnetic oscillations and electrical induction. His instruments, handmade with exceptional precision, drew admiring crowds. Gauss attended Weber's talk and was struck by his experimental skill and clarity of thought. Humboldt's vision of a worldwide magnetic survey had seemed daunting, but in Weber, Gauss recognized the practical partner who might help make it real. Their opportunity came in 1831, when a professorship in physics opened at the University of Göttingen. With Gauss's strong support, Weber was appointed. The partnership was possible in no small part because Alexander's elder brother Wilhelm von Humboldt (1767–1835), architect of the Prussian university reforms two decades earlier, had helped establish the funding streams that supported both research positions and the construction of an astronomical observatory at Göttingen.

Before Gauss and Weber, measurements of magnetic intensity were makeshift. Investigators compared the swing of one needle against another or the angle by which a magnet was deflected, but the results varied with the quirks of each individual instrument and could not be compared across laboratories. Gauss set his sights higher: to measure Earth's field, especially its horizontal component H, in "absolute" terms — expressed only in the fundamental units of mass, length, and time. This was more than a refinement of method. It was a paradigm shift, treating magnetism not as a relative effect but as a physical force whose strength could be captured in universal measures, ready to take its place alongside weight and motion.

To achieve this, Gauss and Weber devised a two-part method using a bar magnet. Every magnet, Gauss reasoned, possesses an intrinsic strength and orientation — its *magnetic moment* (μ). This moment is the magnet's inherent capacity both to generate its own magnetic influence in the surrounding space (its *field*) and to experience a distinct rotational force (a *torque*) when placed within an external field such as Earth's H. The challenge was that neither Earth's field nor the magnet's moment could be measured on its own, since the behavior of a magnet always reflects the interaction of both. Gauss's solution was to design two complementary experiments that, when combined, would disentangle the two quantities.

The first was the oscillation experiment. A bar magnet was suspended horizontally by a fine filament, so it could swing freely like a compass needle. When nudged from its north–south alignment, Earth's field exerted a restoring torque on its μ, causing it to oscillate back and forth. The quickness of these swings, or their period, was timed with great care. This period depended on two different factors: the interaction of μ with Earth's H (the product μH) and the magnet's purely mechanical resistance to rotation, its *moment of inertia*. The *moment of inertia* is not magnetic at all; it depends only on the magnet's mass, length, and shape, and could therefore be calculated directly. From these measurements Gauss and Weber obtained a numerical value for the combined term μH.

The second step was the deflection experiment. Here, the same test magnet was placed at a precisely known distance from a small, freely pivoted compass needle. Orientation was critical: typically the bar magnet was set "end-on", with its axis aligned east–west and passing through the compass center, so that its own field acted at right angles to Earth's north–south direction. The compass needle, possessing its own small magnetic moment, was caught between two influences: the push from the test magnet's field and the restoring pull from Earth's H field. It settled at an angle where the two effects balanced. Measuring this deflection, together with the distance between the magnet and compass, yielded the ratio $\frac{\mu}{H}$. For this purpose Gauss devised the *bifilar magnetometer*, in which a magnet was suspended by two parallel threads of equal length; Weber built and refined the instrument so that even tiny deflections could be measured with precision.

With these two results in hand — the product μH from oscillations and the ratio $\frac{\mu}{H}$ from deflections — Gauss and Weber could solve by straightforward algebra for the individual absolute values of μ and H. The achievement was revolutionary: the strength of Earth's local field and the magnet's intrinsic moment were expressed in terms of mass, length, and time, not tied to the peculiarities of any particular instrument. Although Earth, like any magnet, could also be described by a global *magnetic moment*, this quantity lay outside the reach of their methods. In any case, what mattered to them were the effects observable from the Earth's surface. It was this measurable field, with its variations across geography and time, that became the focus of their wider program.

To extend their absolute measurements into systematic observation, Gauss and Weber set about building Göttingen into a true magnetic observatory. Accuracy depended not only on refined instruments but also on precisely timed readings. Around 1833, to coordinate their observations, the team constructed an electromagnetic telegraph linking the astronomical observatory with the physics institute. Two copper wires, a little over a kilometer long, were strung between the buildings, reportedly running over rooftops and through a church tower. A pulse of current sent along the line caused a magnetic needle at the receiving end to deflect. Because even minuscule deflections had to be detected, they employed an "optical lever": a tiny mirror attached to the galvanometer's needle reflecting a distant scale. By observing this scale through a small stationary telescope in the same room, minute rotations of the needle were amplified into quantifiable shifts. They developed a code based on the direction and number of deflections, enabling them to transmit messages whose chief purpose was the exact synchronization of clocks, so that their sensitive magnetic observations could be tied with precision to astronomical time.

Equipped with these methods, Gauss acted in 1834 to establish the Magnetic Union (Magnetischer Verein), an international network for the study of terrestrial magnetism. Its program was to map the field by meticulously measuring its three principal elements: *declination* (the crucial angle between true and magnetic north), *inclination* (the dip or lift of the magnetic needle away from Earth), and *intensity* (the strength of the field, especially its horizontal component H). The aim was to capture their fluctuations on every scale: the slow "secular" drift that made old charts unreliable, the daily patterns that repeated with the Sun, and sudden disturbances such as magnetic storms. In this way Gauss gave institutional form to Alexander von Humboldt's vision of a collaborative global survey, pursued with the rigor of astronomy.

The effectiveness of the network depended on uniform instruments and methods. Gauss and Weber prepared exact designs and detailed instructions for magnetometers, declinometers, inclinometers, standardized stands, suspension frames, and timing apparatus. More than forty were constructed in Göttingen under Weber's supervision and distributed to observatories across Europe and beyond. The "absolute" character of the method was paramount: instead of relying on transported magnets whose properties might change over time, each observatory could make its own fundamental determinations, repeating the oscillation and deflection procedure to calibrate its own instruments and magnets for accurate measurements.

To ensure comparability, Gauss also issued strict protocols. These specified what was to be measured, how it was to be recorded, and when observations were to take place. A central feature was the designation of "term days", typically four to six each year, including the solstices and equinoxes, when stations carried out twenty-four hours of observations at short, regular intervals, often every five or ten minutes. Additional schedules governed less intensive routine monitoring. The data, recorded in uniform formats, were sent back to Göttingen, where Gauss undertook the immense task of compilation and analysis. The results would appear in the annual volumes *Results from the Observations of the Magnetic Union* (Resultate aus den Beobachtungen des magnetischen Vereins), and become the central reference for the study of terrestrial magnetism.

The spread of observational sites within the Magnetic Union took place through many channels. Existing astronomical and meteorological observatories expanded their programs to include magnetic measurements according to Gauss's system. In other cases, magnetic observatories were founded outright, often encouraged by the influence of Humboldt and Gauss, who solicited support from scientific academies, universities, and governments across Europe. Funding for the stations typically came from local institutions or national treasuries. Staffing was usually provided by astronomers, physicists, or trained observers already attached to these institutions, who, guided by Gauss's detailed instructions, learned to carry out the demanding protocols with precision.

From its German core the network expanded rapidly across Europe: northward to Dublin and Uppsala, southward to Palermo, eastward to Kazan, and into major scientific centers such as Milan, Prague, Stockholm, and Saint Petersburg. The value of coordinated observation quickly became evident. The first returns of data showed that sudden disturbances in compass readings — magnetic storms — occurred at many stations simultaneously. This was decisive evidence that such phenomena were not local curiosities of weather or atmosphere but planetary in scale, affecting Earth's entire magnetic field.

Building on this momentum, Humboldt secured wider backing. He persuaded the Prussian crown to fund improved instruments, gained the crucial patronage of Tsar Nicholas I for a chain of Siberian stations, and in 1836 convinced the Royal Society in London to enlist the British Admiralty. Support came from varied motives: naval offices sought safer navigation from improved charts, mining ministries hoped magnetic fluctuations might hint at ore deposits, and scientific societies looked to fundamental discoveries about Earth's hidden forces. Humboldt's promise of shared data and credit proved essential, persuading rival powers to cooperate in pursuit of a global chart that no single nation could produce on its own.

In November 1837, a political crisis struck closer to home. King Ernst August of Hanover revoked the state's liberal constitution. Wilhelm Weber, co-director of the Göttingen observatory, joined six other professors in a public protest — the Göttingen Seven — and all were dismissed from their posts. Gauss, older and cautious, remained publicly silent, weighed down by family and official obligations, yet he was deeply affected by the loss of his closest collaborator. Weber's removal to Leipzig suspended their daily work together at Göttingen, but by then Gauss already held a vast accumulation of observations: thousands of carefully logged readings from the expanding international network. With the experimental partnership interrupted until Weber's return in 1849, Gauss turned his energies to analyzing this wealth of data, seeking the patterns concealed within Earth's magnetic variations.

The challenge for Gauss was to extract from the growing body of data a coherent picture of Earth's magnetic field as a whole. He began with a simplifying hypothesis: that in the region at and above Earth's surface, where the observations were made, no electrical currents flowed to generate the field. In such a "source-free" space the field could be described by a magnetic *scalar potential*, just as Pierre-Simon Laplace had used to represent gravitation. This potential could be pictured as an invisible landscape surrounding the Earth, its slope at any point corresponding to the direction and strength of the field. As Laplace had shown, such a potential must satisfy his celebrated equation — meaning that in any region without sources the influences spread out smoothly and the potential field has no net curvature. Gauss's assumption was testable: if mistaken, the framework would fail to reproduce the observed measurements. And, importantly, the framework allowed Gauss to ask whether the sources of the field lay primarily within the Earth or outside it.

To construct the global model Gauss turned to spherical harmonic analysis, the same technique Laplace had developed in celestial mechanics. Spherical harmonics can be thought of as a series of mathematical *frequencies* suitable for describing any smoothly varying quantity on the surface of a sphere. By combining the right "amounts" of these frequencies, any smooth pattern can in principle be reproduced. Gauss reasoned that modelling Earth's field did not require the infinite series of all possible frequency terms, but could be approximated well by just the first few. The very first, or lowest-frequency, is the *monopole* term, which corresponds simply to a sphere; the next are the *dipole* terms, each with two opposing lobes oriented differently in space. Gauss set down a system of equations linking magnetic measurements from across the globe to the amounts, or *coefficients*, of these harmonic terms.

Because there were far more measurements than unknowns, and because the data contained inevitable inconsistencies, different subsets of observations would yield slightly different results. To address this, Gauss applied a statistical method he had pioneered: the method of *least squares*. The principle was to treat the "error" as the squared difference between the field predicted by a trial model and the values actually observed, and then adjust the amounts of the harmonic terms to minimize the total error. In this way he distilled the disparate observations into a single mathematical expression for Earth's magnetic potential.

From this analysis Gauss concluded that the great bulk of Earth's magnetism arises from sources deep within the planet. To a close first approximation the field behaves like that of a single, powerful, and slightly tilted dipole. His model quantified both the strength and orientation of this dipole, thereby locating the geomagnetic poles, while also revealing significant regional deviations. Most tellingly, the analysis found no evidence for a monopole contribution: the Earth as a whole carried no detectable magnetic *charge*.

While Gauss remained in Göttingen analyzing the Magnetic Union's data, Weber spent years away, eventually taking a professorship at the University of Leipzig in 1843. Political upheaval soon reshaped the landscape. The Revolutions of 1848 swept across much of Europe, demanding constitutional governance. In Hanover the pressure compelled King Ernst August — the same monarch who had dismissed Weber in 1837 — to reinstate a constitution. In this changed atmosphere the university invited Weber back, and in 1849 he returned to Göttingen, reunited with Gauss for the final years of the latter's life, though their most innovative joint work was behind them.

Weber carried forward the spirit of his collaboration with Gauss: rigorous quantification in search of unifying laws. By the 1840s several experimental facts about electricity and magnetism were well established. *Coulomb's Law* described how stationary charges repel or attract one another with a force that depends on their distance. *Ampère's Law* showed that moving charges exert a different kind of force, magnetic in character, which could not be explained by distance alone. *Faraday's Law* demonstrated that a changing magnetic field, in turn, can induce an electric current. Taken together, these results pointed to an inescapable conclusion: the interaction between charges could not be purely static. Motion itself had to play a role in the force law.

Weber began with Coulomb's inverse-square law as the static baseline. To account for Ampère's magnetic force between moving charges, he introduced an additional term in the force that depended on their relative velocity. But adding such a term raised a problem: velocity carries dimensions of length divided by time, whereas Coulomb's law contains only length. To combine them in a single equation without producing a dimensional mismatch, the velocity term had to be divided by a universal constant that also carried the dimensions of a velocity. This constant was therefore not a free assumption but a logical necessity: any law that unified static and dynamic electricity would require such a fundamental velocity, setting the scale at which magnetic and inductive effects appear when charges move.

In 1856, working with Rudolf Kohlrausch (1809–1858), Weber sought to measure this constant. Their experiment compared the same quantity of electricity measured in two different ways. First, they charged a Leyden jar and measured the electrostatic force it exerted. Then they discharged the jar through a galvanometer and measured the magnetic effect of the resulting current. Dividing the two gave a numerical value for the velocity constant: just over three-hundred-thousand kilometers per second. The number was strikingly close to the known speed of light.

The implication was profound. A constant velocity, demanded on purely theoretical grounds as a conversion factor between static and dynamic electricity, turned out to match the velocity of light itself. The result added new weight to the suspicion, already stirred by Michael Faraday's work on the magneto-optical effect, that electricity, magnetism, and light were different manifestations of the same underlying phenomenon.

Gauss, in his last years, had recognized the depth of these questions. He corresponded with Weber, encouraged the attempt to formulate a single electrodynamic law, and appreciated the possibility that phenomena apparently distinct might share a common foundation. But he died in 1855, before Weber and Kohlrausch's experiment, and never learned that the velocity constant implied by Weber's theory coincided with the speed of light.

While Gauss and Weber's electromagnetic telegraph, by the time of Gauss's death, had stood as a landmark of practical scientific engineering, it was itself built upon earlier experiments in electrical communication. Their Göttingen line was also distinctive in one crucial respect: it was powered by electromagnetic induction rather than by a battery, a design choice that set it apart from contemporary systems. In 1809 Samuel Thomas von Sömmerring (1755–1830), a German physician and anatomist, demonstrated an ingenious electrochemical telegraph. His device used about thirty-five separate wires, one for each letter and numeral, terminating in glass tubes filled with acid at the receiving end. When current from a voltaic pile was sent along a chosen wire, it triggered electrolysis in the corresponding tube, releasing bubbles of gas that visibly marked the selected character. Functional but cumbersome, this multi-wire, electrolysis-based system could not compete with later electromagnetic designs. A generation later Pavel Schilling (1786–1837), a Russian diplomat and inventor, built a more practical *needle telegraph* in Saint Petersburg around 1832. Using the deflection of magnetic needles to represent coded signals, Schilling's design was widely influential and pointed directly toward the technology Gauss and Weber employed.

Russia's contribution to Gauss and Weber's program of mapping terrestrial magnetism was substantial. At its center was Adolph Kupffer (1799–1865), physicist and leading member of the Saint Petersburg Academy of Sciences. Kupffer first met Alexander von Humboldt in Paris in the 1820s and was quickly drawn into his circle of wide-ranging correspondence. Encouraged by Humboldt, and later working closely with Gauss, he became a central organizer of the Magnetic Union. He established a far-reaching chain of magnetic and meteorological observatories across the Russian Empire, from Saint Petersburg deep into Siberia. This network delivered a steady stream of high-quality, standardized measurements from regions that were magnetically important — due both to its broad east-west span and, most crucially, to its northern stations where inclination and intensity changed most sharply. Without this coverage, Gauss's spherical harmonic analysis would have been incomplete.

The same Saint Petersburg environment also produced fundamental advances in electromagnetism. It was here, in 1834, that Emil Lenz formulated his principle that an induced current always opposes the change in magnetic field that produces it.

These efforts built upon a tradition of Russian state patronage of science. Since Peter the Great's founding of the Saint Petersburg Academy of Sciences in 1724, the empire had deliberately recruited European scholars — including Leonhard Euler from the German states — to establish a world-class institution. That tradition of attracting foreign talent and sustaining ambitious research continued into the nineteenth century. Moritz von Jacobi, German-born but working in Saint Petersburg under the patronage of Tsar Nicholas I, exemplified it. In 1834 he developed a powerful rotating electric motor, and in 1838 he demonstrated an improved version by driving a twenty-eight-foot boat with fourteen passengers across the Neva River. That same year he invented *electrotyping* (or *galvanoplastics*), a technique for producing exact metal replicas by depositing copper or other metals onto a conductive mold. The method quickly found use in replicating printing plates for government documents and in the production of sculpture. Jacobi later turned to telegraphy, designing a letter-printing instrument and constructing telegraph lines for the Russian government.

Mikhail Ostrogradsky (1801–1862), a contemporary of Moritz von Jacobi, was educated at Kharkiv University and later in Paris before returning to Saint Petersburg, where he became a leading figure at the Academy of Sciences. By the 1820s many thinkers were converging on a general picture: if an imaginary surface is drawn around a region of space, the total flux of a quantity such as heat, gravity, electricity, or fluid across that surface must correspond to the net amount absorbed by or ejected from the region. Gauss had already demonstrated this balance for certain simple shapes.

While studying heat conduction, Ostrogradsky supplied a rigorous general proof of this idea. He showed that the total flow across a boundary is exactly equivalent to the accumulation of infinitesimal changes inside. In other words, the grand total at the surface is the faithful "shadow" of the countless local variations in the interior. This clarity transformed intuitive balance laws into exact mathematics. His demonstration, later called the *divergence theorem*, was a visible milestone: it revealed the essence of calculus itself — that the infinitesimal and the global are two sides of the same coin, the interior changes inseparably tied to what is measured at the boundary. That insight would become the guiding picture for expressing electromagnetism in precise mathematical form.

The Russian Academy also recognized that scientific prestige was a form of national power, and, like its counterparts in Paris, Berlin, and London, it sought to enhance its standing by electing leading foreign scholars to membership. A prominent example was the election around 1830 of the German mathematician Carl Gustav Jacob Jacobi (1804–1851), younger brother of Moritz von Jacobi. This younger Jacobi was renowned for his mastery of pure mathematics. He developed methods to describe complex periodic motions such as the swing of a pendulum or the rotation of a rigid body — problems that demanded new tools beyond elementary trigonometry. He also established fundamental rules for carrying out calculus consistently across different coordinate systems, such as shifting from a flat grid to a curved surface. Though Jacobi never worked in Russia, election to the Academy gave him both international recognition and access to its publications as a platform for his work. For Russia, honoring a mathematician of Jacobi's caliber signaled that the Saint Petersburg Academy had secured a place in European scientific life, drawing intellectual capital and reinforcing its role in the advancement of fundamental knowledge.

Another foreign member elected to the Saint Petersburg Academy was the German engineer Gustav Kirchhoff (1824–1887). In 1845, while still a student at the University of Königsberg, he formulated two *circuit laws* that became foundational to electrical engineering. His *current law* states that at any junction in a circuit the total current flowing in equals the total flowing out, an expression of conservation of charge. His *voltage law*, rooted in conservation of energy, states that for any closed loop the total voltage gains must equal the total voltage drops. Later in his career Kirchhoff also helped establish *spectroscopy*, demonstrating that each chemical element emits and absorbs a unique "fingerprint" of radiation.

His most profound contribution to electromagnetism came in his 1857 paper *On the propagation of electricity in wires* (Über die Bewegung der Elektricität in Drähten). Kirchhoff imagined a telegraph line as a series of infinitesimal segments, each with three properties: *resistance, capacitance*, and *inductance*. Resistance impeded the flow of current. Capacitance allowed charge to accumulate and an electric field to build. Inductance meant that any change in current created a magnetic field that resisted that change, acting like inertia. Kirchhoff described how these effects combined: when voltage was applied, the first segment charged, current rose, a magnetic field built up, and then as that current slowed, the diminishing magnetic field induced another electric field that pushed the current into the next segment. Energy alternated between electric and magnetic forms, passing forward segment by segment like a wave.

From this reasoning Kirchhoff derived a set of partial differential equations, the *telegraph equations*, showing that in a wire of low resistance the "leapfrogging" behavior obeyed the classical *wave equation*. He further showed that the speed of this electrical wave must equal the ratio of the fundamental electrostatic and electromagnetic units of charge. The conclusion was striking: electrical signals do not act instantaneously but propagate as waves at a definite speed. It was the very same speed Weber and Rudolf Kohlrausch had measured experimentally in 1856 — a value equal to the speed of light.

Kirchhoff's mathematical outlook reflected his training at Königsberg under Franz Ernst Neumann (1798–1895). Neumann was a founder of the German school of "theoretical physics", which sought to capture nature in rigorous equations rather than in Faraday's intuitive descriptions. Faraday had shown that a changing current or magnet could induce a voltage in a nearby wire, but he left it as an experimental observation pictured in "lines of force". In 1845 Neumann set this into exact mathematics in his paper *The General Laws of Induced Electric Currents* (Allgemeine Gesetze der inducirten elektrischen Ströme). He expressed the induced voltage in one circuit as proportional to the rate of change of current in another, with the proportionality fixed by a double integral over the geometries of the two circuits — the quantity that came to be called *mutual inductance*. The result made induction calculable and precise, but it still treated the phenomenon as instantaneous, linking entire circuits through their geometry, rather than describing every point in space. This style of turning experimental facts into formal, predictive relations defined the German approach, and it left a lasting imprint on Kirchhoff's generation.

Meanwhile, across the English Channel, the British scientific tradition was undergoing its own transformation. In the early nineteenth century, British mathematics was still largely in the shadow of Newton, dogmatically adhering to his geometrical methods and cumbersome "fluxional" notation for calculus. This insularity left Britain increasingly isolated from the powerful analytical techniques that were flourishing on the Continent, where Leibniz's compact notation $\frac{dy}{dx}$ had proliferated. After more than a century of relative stagnation, a widespread "climate of reform" began to emerge on several fronts. Britain was finally beginning to shake free and set the stage for the figures who would drive home the electromagnetic revolution.

One of the clearest manifestations of this reform came at Cambridge. Around 1812, a group of rebellious undergraduates — Charles Babbage (1791–1871), George Peacock (1791–1858), and John Herschel (1792–1871) — formed the *Analytical Society*. Their mission, as Babbage later quipped, was to advocate for "the principles of pure D-ism in opposition to the Dot-age of the University", a jab at the notational divide in calculus. The pun captured the essence of the conflict: "D-ism" referred to Leibniz's differential notation, while "dot-age" mocked both Newton's dot notation for fluxions and the perceived intellectual stagnation of Cambridge itself.

Their first step was to translate into English the textbook of the French mathematician Sylvestre François Lacroix (1765–1843), *Treatise on Differential and Integral Calculus* (Traité du calcul différentiel et du calcul intégral, 1797). Lacroix's work was not original research but a pedagogical masterpiece — clear, systematic, and filled with examples that demonstrated the superiority of Leibniz's analytical notation over Newton's fluxions.

The society's next move was a clever subversion of Cambridge's own system. As members became examiners for the all-important Mathematical Tripos, they set questions that were cumbersome to solve with Newtonian methods but straightforward with Lacroix's analytical techniques. This created an undeniable incentive: to succeed in the examinations, students and tutors were forced to adopt the very "D-ism" they had long resisted.

A central ally of the Analytical Society was Edward Bromhead (1789–1855), a close friend of Babbage and Herschel who carried the reforming spirit beyond Cambridge. After returning to his family estate in Lincolnshire in 1812, Bromhead became involved with the nearby intellectual life of Nottingham. At its center stood the subscription library, a private institution where professionals, clergy, and local gentry pooled fees to build a shared collection of books and journals. More than a lending library, it served as the city's scientific club. Guiding its catalogue was John Toplis (1775–1857), a Cambridge-trained mathematician and translator of Pierre-Simon Laplace's advanced works, who ensured that Nottingham had access to texts rarely found in provincial England.

It was into this environment that George Green (1793–1841) emerged. The son of a baker, Green had attended a local academy for only a year before reputedly exhausting what his teachers could offer and returning to work in his father's business. The family's fortunes improved after his father built a successful grain mill — a prosperous enterprise Green would later inherit. This eventual financial independence allowed him to pursue an intense, solitary study of mathematics. In 1823 he joined the Nottingham Subscription Library, gaining access to the continental texts that Toplis had championed. Green's mill at Sneinton stood on the edge of Nottingham, only a short walk from the library, which he frequented with dedication. Within this local intellectual circle, his remarkable, self-taught talent came to light.

In 1828, with the crucial backing of Bromhead and Toplis, Green placed a notice in the newspaper calling for "subscribers" to fund the printing of *An Essay on the Application of Mathematical Analysis to the Theories of Electricity and Magnetism*. Of the fifty-one individuals who pledged support, more than half were fellow members of the library. After the essay appeared in print, Bromhead was both struck by its originality and frustrated by its obscurity. He took it upon himself to promote Green: inviting him to his estate, giving him access to his personal library, and offering the intellectual mentorship and encouragement he had long lacked. In 1831 Bromhead wrote to his friend Charles Babbage, "I think I have found another Laplace in Nottingham; the miller Green is a man of quite extraordinary power in analysis". For years he urged Green to pursue a formal university education, and in 1833 Green finally enrolled at Cambridge, at the age of forty, beginning his academic career.

Meanwhile in Dublin, William Rowan Hamilton (1805–1865) was showing that the British mathematical reform movement was no longer merely catching up with the Continent. Newton had described motion directly in terms of forces and accelerations: a push produces an acceleration set by the object's mass. Lagrangian mechanics reframed this picture by introducing a single function, the *Lagrangian*, which for ordinary mechanical systems is usually kinetic energy minus potential energy. From it the equations of motion can be worked out, with conservation of energy and momentum falling naturally into place.

Hamilton advanced this framework by treating the entire action built from the Lagrangian as the governing principle: out of all the imaginable paths a system could take, the actual one is the one where the action stays "flat" under tiny variations. From this principle he introduced a second central function, the *Hamiltonian*, which in many cases is the total energy. It produces a pair of coupled equations — one for position, one for momentum — that together describe a system's unfolding in time. With Hamilton's reformulation, the universe could not only be described in terms of pushes and pulls, but it could also be seen as a system evolving according to the state and flow of its energy. This energy-centric viewpoint proved to be more fundamental and ultimately more adaptable.

Back at Cambridge, the still relatively unknown George Green was following an independent path into the mathematics of energy and fields. There he befriended Robert Murphy (1806–1843), a brilliant younger mathematician who was one of the few within the academic establishment to recognize the depth of Green's talent. During his productive period in the late 1830s Green published several important papers on waves and optics, though none received much attention in his lifetime. Murphy, however, in one of his own works on the theory of equations, included a passing reference to Green's obscure, privately printed 1828 essay on electricity and magnetism. At the time this single sentence, buried in a specialized text, went almost completely unnoticed by the wider scientific world and did little to elevate Green's profile. Yet it became the one crucial breadcrumb in the academic record that allowed later generations to recover the full, revolutionary scope of his work.

The story of both men ended tragically. Green's health failed; he fell seriously ill in 1840, returned to his family home in Nottingham, and died in 1841. Just two years later, in 1843, Murphy also died, at the young age of thirty-six, nearly taking with him the only surviving trace of Green's masterpiece.

While Green's theoretical work lay dormant, a more pragmatic branch of British science was galvanized into action. In the late 1830s the success of Carl Friedrich Gauss's Magnetic Union initiative prompted Great Britain, with its global naval presence and colonial outposts, to launch its own coordinated survey. This Magnetic Crusade was driven primarily by Colonel Edward Sabine (1788–1883) and John Herschel (1792–1871), who had earlier helped form the Analytical Society. Leveraging observatories in locations as far afield as Toronto, Philadelphia, and Hobart in Tasmania, the Magnetic Crusade produced the densest stream of geomagnetic data yet assembled, significantly advancing the movement toward a comprehensive understanding of Earth's magnetism. The level of international collaboration involved became a model for global science in the generations to come.

By the mid-1840s George Green's 1828 essay was still little known, surviving in only a few rare copies — until it reached a rising prodigy from Scotland, William Thomson (1824–1907). The son of a mathematics professor at the University of Glasgow, Thomson enrolled there at the age of ten before moving on to Cambridge, where he established himself as one of the leading mathematical minds of his generation. He graduated as Second Wrangler in the notoriously difficult Tripos examinations, and in 1846, at only twenty-two, was appointed Professor of Natural Philosophy at Glasgow, a chair he would hold for more than half a century.

Even before that appointment, Thomson had already made a discovery that would shape the course of British physics. In 1845 he was immersing himself in the latest theories of heat and electricity while preparing for a study trip to Paris to work with Europe's leading mathematicians. In a paper by the late Robert Murphy he noticed a footnote: "See Art. 14 of Mr Green's Essay (Nottingham, 1828)". Intrigued by the title and its relevance to his own interests, Thomson began a frantic search for the essay, which was absent from the main libraries. Just before leaving, he appealed to his well-connected tutor William Hopkins (1793–1866) — the famed "wrangler-maker" of Cambridge and a collector of mathematical "quartos". Hopkins, with characteristic foresight, had acquired several of these rare pamphlets directly from Green years earlier.

That spring Thomson packed Green's pamphlets into his luggage and set off for Paris. There, in the study of Joseph Liouville (1809–1882), together with leading figures such as Charles-François Sturm (1803–1855) and Michel Chasles (1793–1880), he laid out the profound results of the unknown English miller. The French analysts, then at the forefront of mathematical physics, were immediately absorbed. As Thomson later recalled, Liouville "gave great attention". Convinced that he had unearthed a lost masterpiece, Thomson arranged for the essay to be republished in Germany's prestigious Crelle's Journal (Journal für die reine und angewandte Mathematik) in three installments between 1850 and 1854, with his own preface urging that every potential theorist should study it.

What so captivated these elite mathematicians was that "everyone found their own work in it". Green's essay contained a powerful unifying insight: a general method of decomposition, breaking down seemingly intractable problems into manageable parts. He was among the earliest in Britain to rigorously adopt and generalize the concept of the *potential function* — a quantity whose gradient gives the force — but his true genius lay in showing how to compute potentials not just in open space, as Laplace and Poisson had done, but in the presence of boundaries and constraints, such as conductors held at a fixed voltage. His celebrated solution for the potential of non-uniform ellipsoids — long considered a nearly impossible problem — demonstrated his approach: reducing complex bodies to simpler concentric shells, layer upon layer, like the skins of an onion. It was a tour de force of method as much as result, and it announced an unmatched level of mathematical sophistication in Britain.

The same theme of decomposition found lasting expression in Green's invention of the method of *Green's functions*. The technique is a universal recipe for solving potential problems by first solving a simpler one. It begins by finding the potential created by a single, idealized *unit impulse* — a point charge — that already respects the specific boundaries of the problem (for example, being zero on the surface of a nearby grounded conductor). This special *impulse response* is the Green's function. Once it is known, the solution for any arrangement of charges within those same boundaries can be obtained by summing, or integrating, the effects of the Green's function for all the sources. It provided a systematic path to a vast class of problems that had previously required bespoke, one-off solutions.

William Thomson recognized this immediately. In his preface to the 1850 republication, and later in his *Reprint of Papers on Electrostatics and Magnetism* (1872), he repeatedly called Green's methods and his famous theorem "indispensable". He argued that Green's work was "the legitimate foundation of every perfect mathematical structure that is to be made from the materials furnished by Coulomb's law".

This early work in potential theory became the basis for Thomson's long and brilliant career, in which he uniquely blended deep theoretical insight with a remarkable talent for solving hard engineering problems. His restless drive produced foundational work on the dynamical theory of heat and, in collaboration with James Joule (1818–1889), the principle of energy conservation, leading to his proposal for an absolute temperature scale, later named *kelvins*. At the same time, he turned to immense practical challenges, most famously serving as scientific advisor and engineer in the effort to lay the first successful trans-Atlantic telegraph cable. His work on signal propagation and his invention of sensitive receiving instruments, such as the mirror galvanometer, were crucial to the project's success. Over his lifetime he patented more than seventy inventions and became a dominant figure in both theoretical and applied science.

Thomson's scientific style was often defined by physical intuition. He would seize on a striking regularity, such as his 1850 insight linking the "circulation" of a fluid around a loop with the accumulated "twist" (or *curl*) within it — the core idea of what is called *Stokes's Theorem* — and present it as a challenge or insight for others to formalize. Content to leave rigorous proofs to his colleagues, he pressed ahead to new territory. For this lifetime of contributions to both fundamental science and industrial innovation, he was ennobled in 1892 as Baron Kelvin of Largs, or Lord Kelvin, the first British scientist elevated to the House of Lords specifically for his scientific achievements.

George Gabriel Stokes (1819–1903) was another giant of the revitalized Cambridge school of mathematical physics. He graduated Senior Wrangler (top mathematics student) in 1841 and in 1849 succeeded Charles Babbage as Lucasian Professor (the same chair once held by Newton), a position he would hold for five decades while also serving as secretary and later president of the Royal Society. His range was vast: he established the law governing the drag on a sphere moving through a viscous fluid, explained the colors of thin films, coined the term *fluorescence* after discovering the blue glow of quinine solution, studied wave motion, optics, and ultraviolet radiation, and pursued the mathematical foundations of hydrodynamics and elasticity. Stokes sought conceptual clarity and analytic rigor in such phenomena, convinced that mathematics could strip away the incidental to reveal governing principles. Deeply religious, he also wrote on the compatibility of science and faith, regarding the search for mathematical order as part of a broader intellectual calling. This blend of rigor, experiment, and pedagogy made him the natural recipient of Thomson's tentative circulation identity — and the one who would shape it into a general theorem of calculus.

Stokes had a remarkably unified vision of calculus. He saw the great integral theorems not as isolated results but as dimensional variants of a single geometric principle: when the derivative of a quantity is accumulated throughout a region, all interior contributions cancel, leaving only a net effect on the region's boundary. On a line this is familiar: all the small rises and falls cancel, leaving only the total change between two endpoints — Newton's fundamental theorem of calculus.

In a two-dimensional plane or three-dimensional space, the principle gives rise to two distinct theorems. The first, concerning divergence, states that the sum of all "sources" within a region equals the total net "outflow" (*flux*) of the field passing perpendicularly through the boundary — the same principle independently developed by Gauss, Ostrogradsky, and Green. Like water poured into a brimming pool must spill over its edge — the divergence theorem says the amount created inside a region always equals what flows out across its perimeter. The second, concerning "curl", states that the sum of all the tiny "swirls" inside a region equals the total circulation along a closed boundary around that region. Like stirring a cup of coffee, the sum of all the eddies spinning within must equal the total sweep around the rim — this was Thomson's revelation.

In 1850, Thomson, likely from his work on the physics of vortices in fluids, discovered this specific relationship between the circulation of a field around a closed loop and the accumulated "twist" (the *curl*) across any surface bounded by that loop. He communicated this result in a letter to Stokes but never published it himself. Stokes, recognizing this as a beautiful example of his own unifying principle, understood its importance immediately. Rather than publishing it himself, he used his influential position as a professor at Cambridge. To test the very best students, he placed the challenge of proving this theorem on the prestigious and notoriously difficult Smith's Prize examination in 1854.

The Smith's Prize examination is a prestigious award for mathematics and theoretical physics at Cambridge University. In the nineteenth century, it was an exceptionally difficult examination taken by the top mathematics graduates (the Wranglers) immediately after their final Tripos exams. While the Tripos tested speed and mastery of the established curriculum, the Smith's Prize was designed to test for originality, physical intuition, and mathematical insight.

One of the young candidates that February was James Clerk Maxwell (1831–1879). Through that exam, Maxwell and an entire generation of Cambridge mathematical physicists encountered the theorem that, thanks to its place on Stokes's paper, became universally known as *Stokes's Theorem*. The two winners of the prize that year — Maxwell and his close friend Edward Routh (1831–1907) — had both been coached by William Hopkins, the same tutor who a decade earlier had provided Thomson with Green's essay and urged his students to study it. Maxwell's complete command of Stokes's Theorem, together with its siblings like the divergence theorem, became an indispensable part of the mathematical toolkit he would wield. With this machinery he translated Michael Faraday's physical concepts of fields and lines of force into an elegant set of partial differential equations — forming a comprehensive theory of electromagnetism.

In the mid-nineteenth century, as Europe converged on a unified mathematical notation, innovation accelerated, and Hermann Ludwig Ferdinand von Helmholtz (1821–1894) emerged as one of its chief drivers. A Prussian army surgeon turned polymath, Helmholtz had a diverse and prolific career. In his essay *On the Conservation of Force* (Über die Erhaltung der Kraft, 1847), he argued on experimental grounds that every physical process must conserve a single quantity: energy. The claim dovetailed with William Rowan Hamilton's earlier mechanics, in which the *Hamiltonian* — generally equal to the total energy — generates motion through stationary action. But Helmholtz carried the idea beyond mathematics, grounding it in measurements from heat engines, galvanic batteries, and muscle work. His authority persuaded Thomson, Stokes, and ultimately Maxwell to treat energy, not force, as the fundamental book-keeping variable. Helmholtz also gave them the right tool: in 1858 he proved that any well-behaved vector field can be split uniquely into a divergent part (from a scalar potential) and a rotational part (from a vector potential) — the *Helmholtz decomposition*. This complementary division gave Maxwell a natural language for the electromagnetic field, much as the division of the plane into horizontal and vertical axes had enabled the analysis of mechanical motion.

Maxwell graduated in January 1854 and stayed on at Trinity College as a Prize Fellow — a paid research position awarded by vote of the Fellows to the most brilliant recent graduates, freeing them from teaching to pursue independent research. During this intensely productive period Maxwell laid the groundwork for his electromagnetic theory.

In February 1855 he began a correspondence with William Thomson. Having studied Thomson's own papers on electricity, Maxwell sought his guidance. He wrote, outlining his ambition to "attack electricity" and asking for a reading list. In his letter Maxwell explained: "I am trying to reconcile your theory of action at a distance, with Faraday's of lines of force… Now I have been thinking… that the best way to get a physical idea of the subject without being committed to any theory is to be looking out for other cases in which the laws are mathematically identical with those of electricity". For Thomson, receiving such a letter from a rising star at his own alma mater was a welcome development.

Thomson's response was pivotal. He had already published papers attempting to place Faraday's ideas on a mathematical basis and recognized the same potential that Maxwell saw. He generously shared his own work and directed Maxwell toward the essential papers of Faraday, André-Marie Ampère, and Siméon Denis Poisson. Thomson also offered the kind of mathematical analogies Maxwell had requested.

His own work on heat flow, for example, employed equations mathematically identical to those governing electrostatics. He urged Maxwell to think in terms of such analogies — to model the behavior of the electromagnetic field without at first being burdened by its ultimate physical nature. Maxwell later recalled: "before I began the study of electricity I resolved to read no mathematics on the subject till I had first read through Faraday's Experimental Researches in Electricity. I was aware that there was supposed to be a difference between Faraday's way of conceiving phenomena and that of the mathematicians... I was first convinced of this by Sir William Thomson, to whose advice and assistance... I owe most of what I have learned on the subject".

Maxwell's fascination with Faraday's writings was profound because he saw value where others did not. To most mathematicians and physicists of the time, Faraday was a brilliant experimenter but a theoretical amateur. Maxwell wrote of Faraday's mind being "stored with a greater variety of images of lines of force than anyone else". He was captivated by the completeness of Faraday's mental picture. He marveled at how a man "who had hardly made a single algebraical calculation" could nonetheless see deeper into electricity than any geometer. Where others saw loose speculation, Maxwell discerned a coherent, local, and dynamic system awaiting its mathematical language. He felt called to provide it. Faraday's "lines of force" were dismissed by most as no more than a useful visualization — a pictorial shorthand for what were believed to be real forces acting at a distance between charged objects. Maxwell saw them differently. To him, the lines of force were not just a representation of the phenomenon; they were the phenomenon itself.

The professional culture of nineteenth-century physics was profoundly shaped by a methodological approach forged by Isaac Newton in the face of a philosophical dilemma. His law of universal gravitation proposed a force that acted instantaneously across the vastness of space, its strength dependent only on mass and distance. The formula contained no variable for time and no mention of any intervening medium. To contemporaries such as Gottfried Wilhelm Leibniz and other mechanist philosophers, this was untenable. They argued that all physical action must occur through direct contact, and that a force acting mysteriously across a void was not a scientific explanation but a return to the "occult qualities" of pre-scientific magic. For a force to be real, they insisted, something tangible had to carry it.

Ironically, Newton privately agreed with the spirit of this criticism. While he defended his law publicly, he confessed in a letter that the idea of one body acting on another "at a distance through a vacuum, without the mediation of anything else... is to me so great an absurdity" that no competent thinker could accept it. He too believed in an intervening agent, but because he could not prove its nature, he rigorously excluded it from his formal theory.

The success of Newton's law of gravitation set a powerful precedent. For over one-hundred-fifty years the theory predicted celestial motions with breathtaking accuracy, all without reference to an intervening medium. While Newton privately believed a deeper mechanical cause must exist, the public success of his work led to a tradition in which the mathematical law itself was often treated as a complete explanation of a physical phenomenon. It was therefore natural that leading Continental physicists of the 1840s — such as André-Marie Ampère, Wilhelm Weber, and Franz Neumann — sought to emulate this success. Following the Newtonian program, they aimed to create an equally elegant, unmediated law for electricity and magnetism, focusing on the mathematical relationships between centers of force.

In this Continental view, potential energy was understood as a property of a system's overall configuration. For a stretched spring, the stored energy belonged to the spring, arising from tension in its structure. Likewise, the energy of an electrical system was considered an abstract quantity of the configuration of charges — a function of their relative positions and motions — not a substance physically present in the space between them. The alternative view, most powerfully championed by Michael Faraday, held that space was not a passive void but was filled with a dynamic medium: the *ether*.

By the mid-nineteenth century, the existence of the ether was widely accepted for an entirely separate reason: the behavior of light. For more than a century, Newton's theory that light was a stream of "corpuscles" had dominated. But this view was overturned by a sequence of experiments. In 1801 Thomas Young (1773–1829) performed his double-slit experiment, showing that light beams could interfere with each other like water waves, producing patterns of reinforcement and cancellation that particles could not explain. Augustin-Jean Fresnel placed the wave theory on a firm mathematical footing. To challenge him, Siméon Denis Poisson calculated that a peculiar consequence of Fresnel's equations was that a bright spot of light should appear at the center of the shadow of a circular object. Poisson presented this as an absurd prediction which would prove the wave theory was wrong. But in 1818 François Arago (1786–1853) carried out the experiment and, to general astonishment, the bright spot appeared exactly as predicted. This was taken as undeniable proof of light's diffraction.

The idea of a wave without a medium was a contradiction: a wave had to be a disturbance of something. The "empty" space between the stars therefore had to be filled with an invisible substance capable of carrying light: the *luminiferous ether*. Its accepted role was to passively transmit the kinetic energy of light waves. Faraday's intuition, and later Maxwell's genius, was to propose that this same ether was also the active agent in all electric and magnetic phenomena.

In his 1847 monograph Hermann von Helmholtz showed that the leading force-based model, Wilhelm Weber's law, was fundamentally incompatible with energy conservation. The formula, if true, would permit energy to be created from nothing — a violation unacceptable to a new generation of physicists, including William Thomson and James Clerk Maxwell. Maxwell seized upon Faraday's concepts of *lines of force* and the *electrotonic state* to make a radical proposal. The ether, he argued, did not only carry the kinetic energy of waves; it could also store potential energy in a static state. The space around a charge was in constant tension or polarization. An electric field was this static stress in the ether, and its energy was physically present throughout the field's volume.

By distributing energy in the tension of the etheric medium, the principle of locality became a necessity. A change at one point could only act directly on its immediate neighbors, so disturbances had to propagate at a finite speed. This contradicted the instantaneous action of older electrical laws and provided the foundation on which Maxwell would build his unified theory of electricity, magnetism, and light.

In December 1855 Maxwell unveiled the first fruits of his labor in his paper *On Faraday's Lines of Force*, declaring his aim was "to show that the science of the electromagnetic field is capable of mathematical treatment without any hypothesis as to the molecular structure of the medium". To achieve this, he devised a physical analogy. He asked his readers to picture an idealized incompressible fluid moving through a resistive medium. This went beyond the simpler "heat flow" analogy, as the mathematics of fluids could represent the swirling vortices characteristic of magnetism, which heat conduction could not. Within this fluid model, he built a complete mathematical system. Electric potential corresponded to pressure in the fluid. The electric field corresponded to the pressure gradient — the force driving the flow. Faraday's lines of force were the velocity and direction of the fluid's motion. Electric charges were sources, where fluid was created, and sinks, where it was destroyed. This ingenious framework allowed him to capture the *kinematics* — the geometry and motion of the field — without speculating on the ether's underlying dynamics.

Maxwell sent a copy of this paper to Faraday, who replied with delight in March 1857: "I was at first almost frightened when I saw such mathematical force made to bear upon the subject, and then wondered to see the subject stand it so well". Later that year Maxwell received another paper from Faraday, titled *On the Conservation of Force*. He wrote back a thoughtful letter, engaging deeply with Faraday's reasoning and gently suggesting a clearer distinction between the modern concepts of *force* and *energy* (or *work*). He also showed how Faraday's idea of lines of force could be applied to gravity.

In November 1857 Faraday wrote again: "having received your last letter, I am exceedingly grateful to you for it… Your letter is to me the first intercommunication on the subject with one of your mode & habit of thinking. It will do me much good; and I shall read and meditate on it again & again". This time he ended with a philosophical plea, voicing his frustration that mathematicians often kept their conclusions locked in a formal language impenetrable to experimentalists. He asked Maxwell: "There is one thing I would be glad to ask you. When a mathematician engaged in investigating physical actions and results has arrived at his own conclusions, may they not be expressed in common language as fully, clearly, and definitely as in mathematical formulae? If so would it not be a great boon to such as I to express them so? — translating them out of their hieroglyphics, that we also might work upon them by experiment. I think it must be so, because I have always found that you could convey to me a perfectly clear idea of your conclusions which, though they may give me no full understanding of the steps of your process, give me the results neither above nor below the truth; — and so clear in character that I can think and work from them. If this be possible would it not be a good thing if mathematicians, writing on these subjects, were to give us their results in this popular, useful, working state, as well as in that which is their own and proper to them[?]".

In a way, Maxwell's career became an answer to Faraday's request. The austere scientific culture he inherited — perfected by the Continental schools — treated physics as a search for the final, minimal mathematical law. Mechanical speculation was regarded as temporary "scaffolding" that ought to be discarded once the formal equations were obtained. Faraday, however, was the greatest living experimentalist and a figure of immense authority. His plea for clear physical intuition over mathematical "hieroglyphics" gave Maxwell license to work differently. A highly visual, geometrical, and mechanical thinker, he relied on physical models as scaffolds to build understanding and to derive equations. With Faraday's moral and intellectual sanction, Maxwell could frame these models not as a weakness but as a deliberate and necessary step towards deeper understanding. He was not merely indulging his own style; he was answering Faraday's call to bridge theory and experiment. This became his unified method of both discovery and explanation.

For the next ten years Maxwell wrestled with the concepts of electromagnetism. His goal was clear: to construct a mechanical theory of the ether that could mathematically account for all known electrical and magnetic phenomena, as Faraday had described them, while also obeying the conservation of energy as Helmholtz had demanded. This ambition immediately raised a practical question: if the ether was a real substance, what kind of substance was it? Was it a rigid solid, a flowing liquid, or a swarm of particles? To answer this Maxwell had to master the principles that govern how different states of matter behave.

Maxwell left his fellowship at Cambridge in 1856 to take up his first professorship at Marischal College in Aberdeen. From there he turned to the Adams Prize, a biennial award at Cambridge established in honor of the astronomer John Couch Adams (1819–1892). The 1857 topic was the unsolved problem of the motions of Saturn's rings. Laplace had shown that a simple, solid ring was unstable, but nobody had yet given a comprehensive theory of what the rings truly were. For Maxwell this presented an ideal test case: a single cosmic system where a solid, a liquid, or a collection of particles were all plausible candidates.

The mechanical framework had been laid out by earlier giants. Leonhard Euler had set out the rigorous equations for fluid dynamics, while Augustin-Louis Cauchy later formalized the mathematics of stress and strain. In this language, *stress* is the force applied to a body, and *strain* is the resulting deformation. These concepts allowed Maxwell to think of matter not just as stuff with mass, but as media that responded in characteristic ways to being pushed, pulled, or twisted. Each classical state of matter can be distinguished by its forms of strain.

A *solid* is rigid. It resists changes in both volume (compression) and shape, and crucially it resists *shear stress* — a force that tries to make its internal layers slide past one another. This shear resistance is what allows a solid to support itself and to transmit vibrations as transverse waves, like the oscillation of a plucked guitar string.

A *liquid* is fluid. Like a solid, it strongly resists changes in volume — it is nearly incompressible. But it offers no resistance to shear stress. Apply a shearing force and it does not spring back; it flows. As a result, a liquid takes the shape of its container.

A *gas* lacks cohesion. It resists neither compression nor shear. In the context of Saturn's rings, Maxwell did not treat it as a pressure-driven gas but as a collection of non-interacting particles. Each particle moved independently, its motion governed only by gravity and momentum, not by forces from its neighbors.

Laplace had tackled the problem by treating a uniform solid ring as a single, rigid body. He hypothetically nudged the entire ring slightly off-center and calculated Saturn's pull on its new position. The result was not a restoring force but one that pushed the ring farther away. This elegantly showed a simple solid ring was unstable, but the method could not address more complex cases, such as if the ring were irregular, flexible, or able to wobble internally.

Maxwell asked a deeper question: not just "where does the whole object go?", but "what happens inside the object?". He treated the ring as a medium through which waves of disturbance could travel, and analyzed their fate. Using *Fourier analysis*, he showed how any distortion — a wobble, ripple, or oval shape — could be broken down into simple waves and tested to see if the ring's own physics would cause them to shrink (stable) or grow (unstable).

The strength of Maxwell's method was its generality. He built a single mathematical framework: an equation of motion for a small segment of the ring that included Saturn's gravity, the ring's self-gravity, and an interchangeable term for the internal forces holding the ring together. By redefining this term — as elastic tension for a solid or fluid pressure for a liquid — he could test each state of matter in turn.

Maxwell began with the solid ring, showing that any *regular*, symmetrical distribution of mass is in a state of impossible balance. A slight displacement creates a fatal positive feedback. The near side is pulled harder than the far, dragging the whole structure further off-center. The instability would feed on itself, creating a self-amplifying wobble that grows exponentially until the ring shatters against the planet. Having proven the general case is doomed, he then used his framework to work backwards, solving for the specific irregular conditions that could possibly be stable. This revealed a single, unlikely exception: a solid ring could survive if it has a massive lump on one side containing eighty-two percent of its total mass.

For a liquid ring he identified a different fatal weakness. The inner edge would orbit Saturn faster than the outer, creating shear forces that would tear the liquid apart, while self-gravity would pull it into clumps. Maxwell showed these forces could never balance. Any ripple in the fluid would be amplified until the smooth stream broke into separate droplets. Unlike the solid case, there were no exceptions: any liquid ring would disintegrate.

Having eliminated solids and liquids, he concluded the only possibility was a vast swarm of unconnected particles. This system is stable against the catastrophic instabilities that doomed the others. Each untethered particle would orbit independently like a tiny moon. A nudge might alter one orbit, but there is no cohesive structure to transmit a destructive wave through the whole system. Collisions would slowly spread the rings over immense timescales, but they would not be destroyed.

Maxwell submitted his essay in 1859. Though its mathematics was dense, the introduction and conclusion were written with striking clarity. He explained the problem and the intuitive meaning of his results with great care. Intriguingly, his was the only submission, but the judges were not obliged to award a prize. Their report was glowing, praising the essay as a work of genius that had solved what was widely considered one of the most difficult open problems in celestial mechanics. Maxwell received £130 for his effort — likely more than half his annual salary.

Maxwell's triumph, however, was immediately followed by an unexpected turn in his personal life. In 1860, while professor at Marischal College in Aberdeen, the city's two ancient universities were merged to create the modern University of Aberdeen. This created redundancies, and in the case of the Professorship of Natural Philosophy, the post went to the other college's professor, who had fifteen years more seniority. The decision was not a reflection of Maxwell's ability but a common administrative solution to redundancy. Though happy and settled in Aberdeen, he suddenly found himself free, and his availability made him a prime candidate for a professorship at King's College London, to which he was appointed that same year. The move placed him at the heart of the scientific world, alongside Michael Faraday. All evidence suggests that it was soon after this move, around 1860, that the two men met in person for the first and only time, when Maxwell visited the Royal Institution. No detailed record survives of the exchange between the sixty-nine-year-old titan of experimental science, whose health and memory were failing, and the twenty-nine-year-old theorist who had embraced the role of mathematical interpreter of the elder's most profound ideas.

Returning to his quest, Maxwell's task was to build a working machine from an incomplete blueprint. His first paper had created a unified mathematical framework for the field's static architecture, synthesizing the work of his predecessors. He incorporated the principle — set out in calculus by Poisson — that electric charges are the sources of the electric field. He also formalized Faraday's experimental conclusion — confirmed by Gauss — that magnetic fields have no sources. What remained was to construct a single system that could also account for the dynamics of electromagnetism. Ampère's law expressed the fact that an electric current produces a swirling magnetic field. But no one had yet combined this with Faraday's observation that a changing magnetic field produces a swirling electric field, into a consistent framework.

Maxwell had assembled the tools to build a detailed mechanical model of the ether — not just as analogy but as hypothesis. This was the purpose of his next paper, *On Physical Lines of Force* (1861–1862), published in four parts. As Faraday had requested, he worked to provide a tangible physical picture alongside the mathematics. His model was a deduction from the physical properties Faraday had already attributed to magnetic lines of force: a tension along the lines, like a stretched rubber band, explaining the attraction between opposite poles; and a mutual pressure perpendicular to the lines, explaining why like poles repel and why the field spreads out to fill space.

Maxwell postulated that a fluid-filled *vortex* was the simplest mechanical system that fit Faraday's two observed conditions. The centrifugal force of the spinning vortex causes it to expand, producing a pressure pushing outward from its axis of rotation.

This accounts for the lateral pressure between magnetic lines of force. At the same time, the dynamics of a vortex make the pressure lowest at its center. This lower pressure along its axis creates an effective tension, drawing it inward and explaining why magnetic lines behave as if under tension.

From this reasoning Maxwell proposed his model: a cellular medium composed of tiny, spinning vortices, with their aligned axes forming Faraday's magnetic field lines. He imagined these vortex cells filling space like a honeycomb. The velocity of each spin corresponds to the intensity of magnetic tension: the faster the spin, the flatter the cell. Between the vortices he placed even smaller *idle wheels*, capable of spinning or moving freely along channels. These wheels allow the vortices to spin uniformly without grinding against one another whenever the magnetic field requires all spins to align.

The wheels also play the role of electric charges. A surplus of wheels in a region represents positive charge; a deficit, negative charge. When the magnetic pressures (vortex spins) on both sides of a wheel are balanced, the wheel simply spins in place, turning in the opposite direction to let its neighbors rotate without clashing. But if one side is under higher pressure (spinning faster), the imbalance makes the wheel roll toward the weaker side. That rolling transfers wheels from crowded regions to sparse ones — a *conduction current* — while also passing angular momentum to the slower vortex and evening out the pressure difference.

Maxwell used the term *electromotive force* for the mechanical shove that appears whenever two neighboring vortices try to turn at different speeds. If one spins faster, it shoves the wheel, which then rolls toward the weaker side. Conversely, if an electromotive force is applied — for instance by connecting a battery — the wheels are pushed through the channels between vortices. Their motion forces the vortices on either side to spin in opposite directions, converting the forward push of the wheels into the magnetic swirl that surrounds a current. In this way the flow of charge is directly linked to the balancing of magnetic pressures.

A key puzzle was the behavior of a capacitor. When a capacitor is connected to a battery, a current flows in the wires while the plates "charge up" with tension, until the current stops. Maxwell needed a mechanism in his ether that naturally reproduced this. He gave each vortex cell an elastic skin, behaving like a torsion spring. When charges are blocked by an insulator, they can temporarily twist the vortices out of position. As in a spring, the restoring force builds until it balances the driving force from the battery, at which point the twisting ceases. If the battery is replaced by a wire, the cell snaps back, reversing the motion in a "discharge".

This winding and unwinding has the same effect in the model as a conduction current: it spins the vortices, however slightly, and induces magnetic pressure.

Maxwell named this frustrated flow the *displacement current*. This insight, reasoned through his mechanical analogy, was pivotal. Although years would pass before he presented it again in final form, Maxwell already saw that he had unified electricity, magnetism, and light in a single coherent theory.

Many British physicists, especially William Thomson, were receptive to Maxwell. They had a strong tradition of physical modeling, thinking in terms of gears, fluids, and tangible mechanisms. They understood and appreciated what Maxwell was trying to do. Many German and French physicists, the heirs of an austere mathematical tradition, were baffled or even disdainful. They saw Maxwell's intricate sea of vortices and wheels as clumsy, unscientific, and even grotesque — a bizarre step backward into metaphysics compared to what they considered their clean and elegant force laws.

The difference reflected two scientific cultures. The Continental tradition valued austere mathematical abstraction. The British school of the Victorian era was rooted in mechanical reasoning. They were children of the Industrial Revolution, surrounded by gears, pistons, and engines. For many of them, a theory was not truly understood until one could imagine building a mechanical model of it. The fact that Maxwell could construct any mechanical system, however contrived, that reproduced all the known laws of electromagnetism, was seen as a triumph. It showed that these phenomena could, in principle, be explained by the familiar laws of matter and motion. Thomson, the most respected physicist in Britain, was the king of model-building. His work was filled with elaborate analogies involving spinning tops, gyroscopes, wax, and jelly to illustrate the properties of matter and the ether. His most famous proposal was the *vortex theory of the atom*, in which atoms were nothing more than tiny, stable, knotted smoke rings in the ether.

The ability to devise a mechanical model was also linked to the demand for energy conservation, which had become the measure of rigor in physics. An abstract force law could conceal violations of conservation, as Hermann von Helmholtz had shown in his critique of Wilhelm Weber's theory. A mechanical model, by contrast, was seen as inherently safer. To Maxwell, a vortex was like a fluid flywheel — a region of ether with a quantifiable amount of kinetic energy in its organized rotation. By proposing a system in which the spinning vortices and moving wheels represented kinetic energy, and the elastic twisting of the ether represented potential energy, he created a complete dynamical system. He could then show that as his "machine" operated, energy was never lost or created, only converted between kinetic and potential accounts. In this way he demonstrated that his theory obeyed the sacred principle of energy conservation.

James Clerk Maxwell's life in London was one of intense and varied scientific activity, much of it undertaken in direct partnership with his wife, Katherine Mary Dewar (circa 1824–1912). The daughter of the Reverend Daniel Dewar, the Principal of Marischal College where Maxwell had held his first professorship, Katherine married James in 1858. When they moved to London for his new post at King's College, she became an indispensable part of his professional life. Though not a trained scientist, she was intelligent and meticulous, acting as his trusted lab assistant. She spent hours with him in their top-floor London flat, helping with delicate and tedious experiments measuring the viscosity of gases and, most famously, serving as the critical observer in his groundbreaking work on human color perception. This was no minor side project but a fundamental investigation into perception itself. Building on the *trichromatic theory* of Thomas Young (1773–1829), their key result was to prove that any color could be created by mixing specific amounts of just three primary colors of light: red, green, and blue.

Their collaboration produced one of the most stunning scientific demonstrations of the era. In a landmark 1861 lecture at the Royal Institution, in the same building where Faraday had worked, Maxwell unveiled a projected image of a colorful tartan ribbon — a startling achievement. The methodology, devised by Maxwell and executed by the pioneering photographer Thomas Sutton (circa 1819–1875), was a direct demonstration of the color theory. Sutton took three separate black-and-white photographs of the ribbon, each with a different colored filter placed in front of the lens: one red, one green, and one blue. During the lecture, Maxwell used three "magic lanterns" (projectors). He placed each black-and-white slide into its own projector, fitted with the same colored filter used to take the original photograph. When the three filtered projections were perfectly aligned on a screen, the red, green, and blue images recombined into a single full-color reproduction of the ribbon. The demonstration established the foundational principle of additive color photography.

During this same period, Maxwell's growing reputation as a theorist led to a task ideally suited to his talent for connecting abstract mathematics with precise measurement. In the 1860s the science of electricity was a chaotic frontier with no agreed standards, a problem made urgent by repeated and costly failures of submarine telegraph cables. To address it, the British Association for the Advancement of Science formed a committee of scientific luminaries, including Maxwell and William Thomson. Maxwell showed, from first principles, that electrical resistance could be defined in terms of mass, length, and time — in fact, it had the dimensions of a velocity. This insight guided the committee's painstaking experimental work to create the first standard *Ohm*: a physical coil of wire precisely calibrated to embody that definition.

Having captured the core dynamic relationships of electromagnetism by 1862, Maxwell spent the next few years searching for a better way to present his theory. The culmination of this work came with his paper, *A Dynamical Theory of the Electromagnetic Field* (1865). Here, having used a mechanical model to make his great discovery, he finally let the scaffolding fall away. Instead of deriving his equations from a hypothetical machine, he presented the abstract mathematical relations themselves as the fundamental axioms of the theory. This shift elevated the laws of the field to be the primary description of reality, independent of any specific mechanical implementation.

In this paper Maxwell presented his work for the first time as a single coherent system. He summarized it as a list of twenty scalar equations in twenty variables, organized neatly into eight lettered groups (A through H). It was the systematic presentation of a complete theory of electromagnetism. His treatment of energy was also more sophisticated. He moved beyond using conservation as a mere constraint, and developed equations that described the density and flow of energy within the field itself.

The 1865 paper restated Maxwell's most world-changing insight, which he had already reached three years earlier but few had followed through his mechanical model. Faraday had shown that a changing magnetic field induces a swirling electric field. By adding the displacement current term, Maxwell discovered the reverse: a changing electric field creates a swirling magnetic field. In other words, a magnetic field appears not only around a steady electric current, as Ampère's law had stated, but also wherever the electric field is changing.

This necessarily leads to a self-perpetuating cycle. A change in the magnetic field begets a change in the electric field, which begets another change in the magnetic field, and so on — a leapfrogging chain reaction through space. Such a self-sustaining disturbance is, by definition, a wave. The mathematics were the same as Gustav Kirchhoff had shown for electrical signals in a perfect telegraph wire, but Maxwell realized no wire was needed. Using his completed field equations, he too calculated the speed of these electromagnetic waves in the ether, confirmed it equalled the observed speed of light, and wrote: "The agreement of the results seems to show that light and magnetism are affections of the same substance, and that light is an electromagnetic disturbance propagated through the field according to electromagnetic laws".

Between finishing the manuscript in the autumn of 1864 and publishing the paper in June 1865, Maxwell resigned his chair at King's College London. He cited poor health and a wish for the quiet he could not find in the city. He retired to his family estate in Glenlair, Scotland. This apparent retreat, however, was a strategic withdrawal that unleashed a burst of creativity. In the years that followed he consolidated the foundations of two separate branches of modern physics. His days were disciplined; his letters suggest he spent mornings on his studies of heat and gases, and afternoons on the monumental task of writing his magnum opus on electricity and magnetism.

His work on gases grew out of two papers from 1860 and 1867. In them he asked readers to imagine a gas as an immense crowd of identical, perfectly elastic marbles flying in all directions. The constant hail of tiny kicks these marbles deliver to any surface perfectly explains gas pressure. By working out the statistical distribution of their speeds — the bell-shaped *Maxwell distribution* — he demonstrated that temperature is nothing more than the average energy of this ceaseless jostling. The same model also explained an experimental puzzle: why the viscosity of a gas does not increase as it becomes denser.

Maxwell then used these results to probe how heat can leap across the empty gap between the Sun and the Earth. If heat is simply the motion of molecules, it cannot travel through the vacuum of space where no molecules exist to move. Guided by energy conservation, he reasoned that the warmth leaving the Sun must persist in the intervening space as energy stored in an unseen medium. Space itself, he argued, must behave like his elastic ether, capable of carrying energy as a traveling ripple of radiant heat.

In 1800 William Herschel (1738–1822) had spread sunlight through a glass prism, slid a thermometer past the red edge, and found the strongest warming effect just outside the visible band. Because a prism bends longer waves less than shorter ones, he concluded the warm, unseen region must be made of waves with longer wavelength — and therefore lower frequency — than red light. Later experiments confirmed that these invisible waves could be reflected, focused, and polarized just like ordinary light.

Maxwell gathered these insights in his textbook *Theory of Heat* (1871). The Sun's heat, he explained, must shake the surrounding ether, creating vibrations across a wide range of frequencies. Some faster vibrations appear as colored light, while slower ones are felt as radiant heat. He concluded there was no need for a special *caloric* fluid, a still-popular theory at the time. Radiant heat was simply another form of light — an invisible color in the continuous electromagnetic spectrum, governed by the same field equations and traveling at the same speed.

As Maxwell was developing his theories of gases and the ether — both systems composed of countless interacting parts — he confronted the universal problem of stability. In an unstable system, a small disturbance can make the whole arrangement fly apart. In an apparently unrelated paper of 1868, *On Governors*, he analyzed the "fly-ball governors" of steam engines and in doing so created a rigorous mathematical theory of feedback and self-regulation. He translated the mechanical system into differential equations and derived a *characteristic equation* whose mathematical solutions, or *roots*, determined stability. If all the roots had negative real parts, disturbances would quickly decay and the system was stable. If any root had a positive real part, the system was unstable, leading to runaway oscillations.

His analysis showed how a poorly designed governor created a disastrous feedback loop. A small speed-up would make the arms overshoot and slam the steam valve too far shut, bogging the engine down. The arms would then fall and overshoot again, yanking the valve wide open and flooding the engine with steam, causing it to race to an even higher speed. The self-amplifying cycle, fed by the engine's own power, could tear the mechanism apart. By identifying the parameters that led to this condition, Maxwell provided a blueprint for engineering a stable feedback system.

In 1869 George Gabriel Stokes chaired a Cambridge Senate committee that confirmed what lecturers had long complained: the university taught superb mathematics but had no proper experimental laboratory for physics. The report concluded it would cost £6,300 to establish a new chair and laboratory. At almost the same moment, sixty miles north in Glenlair, Maxwell was revising his manuscripts on heat and electromagnetism and writing polite but insistent letters about an untapped scientific treasure: Henry Cavendish's unpublished notebooks on electrical capacity.

Working in total seclusion a century earlier, Henry Cavendish (1731–1810) had carried out measurements of stunning precision on *capacitance* and *dielectric constants*. He built ingenious devices to measure the capacity of different capacitors and systematically tested how inserting various insulating materials — glass, wax, shellac — changed their ability to store charge. In effect he had discovered the concept of *permittivity*, a cornerstone of Maxwell's theory, decades before it was understood. The existence of Cavendish's trove of unpublished papers was known within elite circles. Maxwell learned of their importance through his Cambridge connections, particularly William Thomson. He knew the data existed, but not exactly what it contained. What he did know was that his own theory made precise predictions about the role of insulators in storing electric energy, and that Cavendish's legendary precision might provide both a historical precedent and an experimental foundation for his ideas.

The missing piece appeared in October 1870, when William Cavendish, seventh Duke of Devonshire — Cambridge Chancellor and Cavendish heir — unexpectedly pledged the exact £6,300 to endow a new professorship and laboratory bearing his ancestor's name. The electors, observing etiquette, first offered it to William Thomson and then to Hermann von Helmholtz, who both declined. Helmholtz by then was perhaps the most celebrated scientist in the world. His fame rested on a vast portfolio of universally recognized achievements: he had established the conservation of energy as a fundamental law, invented the ophthalmoscope, and produced landmark studies of sight and sound. In comparison, Maxwell was known to the elite as a brilliant mathematical physicist, but his electromagnetic theory was still seen as abstract, difficult, and unproven. By Victorian academic custom, the ritual of offering the chair elsewhere first was less about genuine preference than about deference to reputation and etiquette, even though the offer seemed custom-built for persuading Maxwell back to Cambridge.

During his semi-retirement Maxwell maintained a strong connection to Cambridge, traveling down each year from Glenlair to examine candidates for the Mathematical Tripos. These visits gave his friends opportunities to cultivate him for the role they were creating. Their days fell into a comfortable rhythm of formal dinners and, more importantly, long walks along the "Backs" by the River Cam. It was in these conversations that they appealed to Maxwell's sense of duty to shape the future of British science. When the opportunity availed itself, almost overnight, Stokes, Arthur Cayley (1821–1895), and the young John William Strutt (1842–1919) — heir to the title Lord Rayleigh — began a friendly barrage of letters offering Maxwell a chair, a laboratory, and Cavendish's long-sought papers. Maxwell was reluctant. He cherished his quiet life of research at Glenlair and initially declined to be a candidate. But in late February 1871 he relented, and two weeks later he was appointed the first Cavendish Professor.

Maxwell returned to Cambridge that autumn. His formal duties were to design and oversee the construction of the new laboratory, and later to lecture and perform research, but everyone understood that his great project — finishing the Treatise — was the priority. The university had effectively given him the time and resources to polish the book that would define the research program of the laboratory he was building. Over the next eighteen months he divided his days between the builders, the university archives where he was studying Henry Cavendish's long-lost manuscripts, and the printers, to whom he was sending final proofs. The two projects came to fruition together: the Cavendish Laboratory opened its doors in June 1873, and one month later the completed *Treatise on Electricity and Magnetism* appeared, becoming the foundational text for the science that would be explored within its walls. In this definitive work Maxwell reviewed all known experimental data, compared his own *field* theory to the rival *action-at-a-distance* theories of Wilhelm Weber and others, and unified it all within a formal mathematical structure. He filled the book with physical explanations and mechanical analogies alongside rigorous mathematics, creating a systematic derivation of electromagnetic theory from first principles.

Published six years after Michael Faraday's death, the Treatise was also a tribute to him. It was written in the spirit Faraday had requested in his letters: mathematical results expressed with physical clarity. Though Faraday's failing health had prevented further meetings, Maxwell continued to send him papers out of respect, even when the elder scientist could no longer follow them. The Treatise was built around Faraday's concepts of lines of force and the physical field, demonstrating with rigor the core truth of what Faraday had seen in his mind's eye.

Maxwell's achievement was to see how a collection of disparate observations and facts about aggregate behavior could be stitched together into a single, precise model of electromagnetism. The experimental laws that Faraday, Ampère, and others had written down described totals — how much force was gained around an entire circuit, or how much field passed through a whole surface. Maxwell recognized that the laws linked two *orthogonal* actions — a field circulating around a boundary and another changing across the surface it enclosed. If this relation held for every circuit, then it must hold everywhere in space. Calculus provided the means to show the transformation explicitly, turning these global loop and surface statements into local differential equations.

Faraday had shown that the total current-driving force — the voltage — around a closed loop is proportional to the rate of change of the *magnetic flux* through the surface it encloses. Maxwell took the first step by rewriting this total *electromotive force*, using the topological principles he had learned from Stokes. He knew that a circulation measured around a boundary can equally be written as a sum of the local "rotation" at each interior point. With this in hand, the changing magnetic flux could likewise be expressed as a sum of pointwise changes of the magnetic field across the surface. And because the equality held whatever surface one chose to span the loop, the only way the two sides could always agree was if their contributions matched point by point. Dropping the totals, Maxwell stated the local law: the rotation of the electric force at each point in space equals the negative rate of change of the magnetic field.

Maxwell applied the same logic to Ampère's law, a form of inverse to Faraday's: that the total magnetic force around a closed loop is proportional to the total current passing through the surface it encloses. This was the case where Maxwell had added a novel term to the equations he inherited, completing the law by including the *displacement current* — the rate of change of the electric field — alongside the usual conduction current. With that amendment in place, the same calculus transformation could be applied: the loop-total of magnetic force could be written as the surface-total of local rotation, and the equality of totals implied that their pointwise contributors must match everywhere. The local law then followed: at each point in space, the rotation of the magnetic force equals the current.

Maxwell treated the Gauss laws accordingly. In their experimental form they stated that the total electric flux through a closed surface is equal to the total charge enclosed within it, and that the total magnetic flux through any closed surface is zero. These were again aggregate statements: surface-totals compared with sums of sources inside. By the same topological relation, each total could be rewritten as the sum of the field's local *divergence* emanating from each interior point. And because the equality held for every closed surface, the only way the totals could always agree was if their contributions matched point by point. From this Maxwell stated the local conditions: at every point in space the divergence of the electric field equals the charge density, and the divergence of the magnetic field is zero. Although he would refine the notation in his final Treatise, he presciently remarked that "the mathematical shorthand of the future will no doubt condense them" — seeing the relations clearly but lacking the symbols to compress them further.

Maxwell fell ill in 1877 and died in 1879, at the age of forty-eight. The illness, later recognized as abdominal cancer, was the same disease that had taken his mother at a similar age. He never lived to see the experimental confirmation of the displacement current or the waves his equations had foretold. The mathematics was in place, but the technology to test it was not. Proof came eight years after his death, in 1887, in the Karlsruhe laboratory of Heinrich Hertz (1857–1894). Using an oscillating circuit and a spark gap, Hertz generated electromagnetic waves, sent them across the room, and detected them with a simple receiving antenna. He showed that these invisible waves behaved exactly as Maxwell's equations predicted, traveling at the speed of light, and exhibiting reflection and refraction. Hertz's experiment not only provided a compelling confirmation of Maxwell's theory, but also ushered in the age of radio. Knowledge and power had never been so entwined.

Ripples Through Reality

The completion of Maxwell's electromagnetic theory was not an end, but a detonation. The elegant equations, describing the intricate dance of invisible fields, held within them the seeds of a new world. In the wake of this triumph, the mathematical framework Maxwell originally employed was refined and condensed by another generation of mathematical physicists. Like Prometheus, these great synthesizers delivered a divine fire to humanity: the secret of light itself. It was a gift of limitless potential, promising a new age of effortless global communication and boundless energy. Zeus's punishment for stealing from Olympia was two-fold: Prometheus, the giver of knowledge, suffered eternal torment, while the recipients, blinded by Pandora's beauty, were plagued by untold evils lurking within her box. So too would it be for humanity in this next age.

What had begun as a battle over the telegraph wires connecting continents, shifted focus to the electrical grids that powered cities. The ether itself was contested next, in a chaotic race to own the medium of global communication. The commercial battles inevitably escalated, until the rivalry of corporations became the strategy of nations, and power was ultimately consolidated by the absolute authority of nation states at war. The decisive end of this tumultuous era was marked by the cataclysm of The Great War. The quest for a more perfect understanding of nature could not compete with the demand to control the power that had already been unleashed.

The man who would most profoundly shape the legacy of Maxwell's work was not a Cambridge academic, but a reclusive, self-taught engineer named Oliver Heaviside (1850–1925). Heaviside was born into a modest London family, partially deaf from a bout of scarlet fever during his childhood. His father, Thomas Heaviside (1813–1896), was a skilled wood engraver. This was a highly respected craft in the mid-nineteenth century, essential for illustrating books and newspapers. However, during Oliver's youth, this profession was being systematically wiped away by a new technology: photography, and the accompanying process of photoengraving. Heaviside grew up watching his father's intricate, manual craft become obsolete.

In 1861, when Oliver Heaviside was eleven, his older brother Arthur left home to begin a promising career, taking a post at a telegraph company station in Newcastle. The position was secured through the immense influence of their uncle, Charles Wheatstone (1802–1875), a primary architect of the telegraph industry. In an era where a university education was a rare privilege, a career in telegraphy was one of the most prized technical professions one could aspire to. At the age of sixteen, Oliver's formal schooling came to an end, and he followed his brother's path north, working alongside him for a time to learn his trade. In 1868, he took a position of his own, working on the international undersea telegraph cables — a challenging and exciting high-tech job.

Heaviside's uncle by marriage, Charles Wheatstone, was an English physicist and inventor who, alongside William Fothergill Cooke (1806–1879), pioneered electric telegraphy in Britain. He developed the *Wheatstone bridge* — a simple but powerful circuit that determines an unknown resistance by balancing it against a set of known ones, making it possible to detect even the smallest variations with great precision — and, with Cooke, patented a needle-based electric telegraph in 1837. Beyond his inventions, Wheatstone was an influential authority within the rapidly growing telegraph industry. The early telegraphs of the 1830s, like the one built by Carl Friedrich Gauss and Wilhelm Weber for their Magnetic Union, proved that messages could be sent electrically, but they were not yet capable of spanning a nation. Cooke and Wheatstone transformed the telegraph from a scientific instrument into an industrial technology.

In 1836, while attending a university lecture in Heidelberg, Cooke witnessed a demonstration by his professor of a primitive needle telegraph based on a design by the Russian diplomat Pavel Schilling. Though he had been studying to create detailed anatomical models, Cooke was instantly seized by the telegraph's immense commercial potential and abandoned his studies. Lacking the deep scientific expertise to perfect it, he returned to England and sought out the advice of Michael Faraday. Faraday referred Cooke to the secretary of the Royal Society, Peter Mark Roget (1779–1869), who facilitated a formal introduction to the already famous inventor Charles Wheatstone. Together, they completed the invention of a five-needle telegraph system, which they filed for patent in June of 1837. Within two days, they had secured provisional protection.

With no direct precedent to draw upon, Cooke and Wheatstone had to prove both the technical feasibility and commercial viability of their invention. The railways had not only a need for, but the physical infrastructure to support such a long-distance signaling system. As railway lines got longer and busier, preventing collisions at junctions and coordinating traffic through long tunnels became a critical safety and efficiency problem. In addition, the railroad companies exclusively owned long, continuous, and relatively secure strips of land — the most practical path for stringing miles of uninterrupted wire. In 1838, Britain was in the grip of its first "railway mania" — a period of explosive, chaotic, and speculative growth. There were not just a few companies, but dozens of railway companies being formed, all competing for investment and parliamentary approval to build lines. Cooke, the tireless entrepreneur, targeted the most ambitious project in England: the Great Western Railway.

Not just another railway, the Great Western Railway was conceived as a masterpiece of modern engineering; a high-speed line designed to be the finest in the world. The visionary behind this immense undertaking was its chief engineer, a man already considered a giant of the industrial age: Isambard Kingdom Brunel (1806–1859). Brunel was legendary for his audacious vision and his fanatical attention to detail, personally designing everything from the revolutionary "broad gauge" tracks to magnificent bridges and tunnels. When Cooke approached Brunel in 1838, he was approaching a man who was simultaneously famous for surviving the world's most dangerous tunnel, designing the world's longest bridge, building the world's most advanced railway, and launching the world's largest ship. Brunel was intrigued by Cooke's proposal to solve the problem of managing train traffic along his busy, high-speed lines. After a period of relentless lobbying and negotiation, Brunel sanctioned a thirteen mile trial telegraph line from London's Paddington Station to West Drayton. Cooke would be permitted to construct on the railway's land, in exchange for providing signaling capabilities to the railroad operators. By 1839, the initial line had been completed.

From a technical standpoint, the pilot program was an unqualified success. The instruments worked flawlessly, messages were transmitted instantly, and Isambard Kingdom Brunel was reportedly deeply impressed with the technology's potential. However, the Great Western Railway's board of directors, focused on the immense cost of building the railway itself, viewed the telegraph as an expensive and unproven novelty rather than an essential system. They declined to fund its expansion, leaving Cooke and Wheatstone in a precarious position: they had shown that their invention worked, but they had not yet demonstrated that a wider, paying market for it existed.

Cooke refused to let his invention suffer the same fate as that of Francis Ronalds (1788–1873). Decades earlier, in 1816, Ronalds had demonstrated two working electric telegraphs. One circuit was buried underground in a trench, and the other strung out as eight miles of wire in his mother's garden. He offered his invention to the British Admiralty, only to have it infamously dismissed in writing as "telegraphs of any kind are now wholly unnecessary". Cooke, facing similar institutional indifference from the Great Western Railway's board, made a bold gamble. In 1843, at his own expense and considerable personal risk, he funded the extension of the line another five miles to the high-profile town of Slough. His goal was to reach the general public by opening the world's first telegraph office where anyone could pay a shilling to see the marvel and send a message. It was this extension that led to the capture of John Tawell in 1845.

The 1845 capture of the murderer John Tawell was the single event that cemented the telegraph's place in the public imagination in Britain. Tawell had poisoned his mistress and boarded a train from Slough to London to escape. A description of him was wired ahead to London using the Cooke-Wheatstone telegraph along the Great Western Railway line. When Tawell arrived at Paddington Station, thinking he was safe, police were waiting to tail him, leading to his eventual arrest and execution. The story became a media sensation. Newspapers dubbed the telegraph "the cords that hung John Tawell". For the first time, the public saw the technology not as a scientific curiosity or a business tool, but as a miraculous instrument of justice that could move faster than a speeding train. It was a killer app for public perception, demonstrating its immense power in a way everyone could understand.

For John Lewis Ricardo (1812–1862), a shrewd Member of Parliament and chairman of another major railway, this was the final proof of concept needed. He saw beyond their existing contracts for railway signaling, and envisioned a privately owned network that could transmit stock prices, political intelligence, and press dispatches at the speed of light, generating immense profits for whoever controlled the wires. Recognizing that the moment was ripe for a massive investment, Ricardo brought his considerable capital and political influence to the struggling partnership of Cooke and Wheatstone. In 1846, he led the formation of the Electric Telegraph Company, the formidable enterprise that would become a powerful industrial monopoly. The country had moved from "railway mania" to "telegraph mania". By this time, Cooke and Wheatstone's partnership, fractured by years of public feuding over credit, had become untenable. Wheatstone, wishing to return to his intellectual pursuits, negotiated a lucrative settlement of cash and company shares for his patent rights. Cooke was then free to pursue the corporate venture with Ricardo, though he would remain tragically obsessed with his campaign for credit.

In 1838, just as William Fothergill Cooke was engaged in his crucial lobbying of the Great Western Railway, an American rival arrived in London. Samuel Finley Breese Morse (1791–1872), a painter by trade, was on a mission to secure British patents for a telegraph system he was presenting as his own singular invention. Morse had already filed for a patent caveat in the United States, but believed that whichever system dominated the British Empire would dominate the world. Over several weeks, Morse conducted a series of private demonstrations for the scientific elite of London, including Michael Faraday and Charles Wheatstone, attempting to establish the priority and superiority of his system. He held separate business meetings with Cooke, which effectively served to confirm their rival status. His patents would be rejected in the United Kingdom on the grounds of Cooke and Wheatstone's legal priority.

After being blocked in London, Morse continued his roadshow to Paris later in 1838, hoping for a better outcome. The result was another successful failure. He demonstrated his telegraph to the prestigious French Academy of Sciences. The influential physicist and politician François Arago wrote a glowing report praising the ingenuity and utility of Morse's device. The French government, however, held a strict state monopoly over all forms of national communication. They had invested heavily in their extensive optical *semaphore* telegraph network — a system of towers with large mechanical arms that relayed messages visually across the country, with the help of telescopes. They were simply not interested in funding or permitting a competing electrical system that would render their established state-run network obsolete.

In truth, the apparatus Morse demonstrated was a synthesis of the work of others. The fundamental science that made it possible — the powerful electromagnet and the crucial concept of the electrical relay — had been discovered and freely shared with him by the American physicist Joseph Henry. Furthermore, the practical, robust machine Morse carried with him was largely the creation of Alfred Vail (1807–1859), a gifted machinist who had transformed Morse's crude prototype into a working system, inventing the simple sending key, the embossing paper tape receiver that made it functional, as well as the familiar coding system of dots and dashes. This partnership was possible because Vail's family not only provided the crucial funding for the venture but also gave Morse the full use of their sophisticated Speedwell Iron Works. Upon his return from Europe, Morse would also adopt a single wire with ground return for his system, a major innovation pioneered by the German engineer Carl August von Steinheil (1801–1870).

Returning to America in 1839, Samuel Morse faced an American landscape vastly different from the one his British rivals enjoyed. The American railways, far from being the dense and profitable network of Britain, were sparse, disconnected, and fighting for survival. With no capital of his own, nor private financiers to appeal to, and no enthusiastic industrial partners to lobby, Morse saw only one, uncertain path: to convince a skeptical and frugal United States Congress that his invention was a matter of national importance. While Cooke had needed to prove both the viability of his product and the existence of a market, Morse was forced to confront a more systemic economic issue.

In the mid-1830s, an intoxicating fever of speculation had gripped the United States. Fortunes were being made overnight in western land deals and a booming cotton market. It seemed to many that the path to wealth was simply to borrow and buy. The fuel for this fire was cheap and plentiful credit from the Bank of England pouring into the young, high-growth American economy. But in London, the governors of the Bank of England watched this speculative frenzy with growing alarm as their own gold reserves drained west to fuel the American bubble. In 1836, they acted with decisive force. Without warning, they sharply raised interest rates, effectively cutting off the flow of capital to the United States. The effect on the American economy was catastrophic. The shockwave, which would become known as the Panic of 1837, collapsed banks, wiped out fortunes, and plunged the nation into a deep and lasting depression.

The strategy to win the backing of the United States Congress had been set in motion well before Morse ever set sail for Europe, when he formed a crucial but fraught partnership with the shrewd chairman of the House Committee on Commerce, Francis Ormand Jonathan Smith. Smith, recognizing the immense value of Morse's patents, had leveraged his position to become a legal partner, demanding a massive stake in the enterprise in exchange for his political muscle — a deal that came at the expense of Morse's original engineering partner, Alfred Vail. For years after Smith left office in 1839, he and Morse waged their campaign from the outside, using the results of the European tour as their primary weapon. They presented the praise from French academics as proof of American genius, while using the existence of the British telegraph as a competitive threat to stoke national pride and fear. They framed the invention as both a vital tool for national defense and a high-speed extension of the government's own postal service to sidestep constitutional objections.

Their efforts culminated in early 1843 with a pivotal demonstration inside the United States Capitol, where messages sent between two committee rooms made the technology's reality undeniable to the lawmakers. This piece of political theater gave the bill's primary champion on the House floor, the Congressman John Pendleton Kennedy, the evidence he needed to counter the opposition's ridicule. The bill also had a key supporter in the Senate, Robert James Walker, who steered it through the more deliberative and skeptical chamber. On the last night of the legislative session, March 3, 1843, while Smith directed the strategy from the outside, Kennedy managed the house floor and pushed the bill to a vote, amidst a chaotic mix of everything from major appropriations bills to minor private land claims. The bill was then rushed to the Senate, where Walker ensured its passage. It was immediately sent to the White House, where President John Tyler, preoccupied with larger matters like the annexation of Texas, and seeing no reason to veto, signed it into law. It passed by a thin margin — a victory secured by the combined force of Smith's years of relentless lobbying and their team's deft parliamentary maneuvering. The Telegraph Bill was a small, strange item in a massive, rushed legislative docket.

The Telegraph Bill appropriated $30,000 to construct and test a forty mile line between Washington D.C. and Baltimore. To secure the route, Morse struck a deal with the Baltimore & Ohio Railroad, which granted him the right-of-way to run his wires alongside their tracks in exchange for the future use of the service. Morse's plan was to insulate the copper wire and encase it in a lead pipe, then bury it underground, alongside the railroad track. However, this was not something he had tried before.

Ezra Cornell (1807-1874) was a pragmatic and ingenious Quaker mechanic from upstate New York. Cornell was in Maine in August of 1843, demonstrating the clever horse-drawn plow he had invented to dig a trench and lay pipe simultaneously, hoping to sell licenses to his patent. Francis Smith, who was struggling with the problem of how to bury the lead pipe for the experimental line, by chance witnessed one of these demonstrations, and immediately saw that Cornell's machine was the solution his team was looking for. Smith hired Cornell on the spot to oversee the task of laying the wire. After returning home to Ithaca to get his affairs in order, Cornell arrived at the construction site outside Baltimore in October. He was immediately alarmed. Watching the process, he suspected that pulling the delicate, insulated wire through the hot, newly-formed lead pipe was causing damage. He insisted on halting the work to test the already-laid sections. His suspicions were correct; the line was electrically dead. With the project in crisis and a third of the budget already spent on a failed method, Cornell sequestered himself for the winter in the libraries of Washington D.C., studying every European engineering text he could find. He returned to Morse not just with a diagnosis, but with a plan: abandon the pipe and string the wires on overhead poles using a glass insulator of his own design. He then re-negotiated his role from a salaried supervisor to a contractor for a fixed fee, confident he could do the job more efficiently than anyone else. Cornell's crew began work in the spring of 1844 and completed the entire forty mile line in just a few months.

On May 24, 1844, from the chamber of the Supreme Court in the Capitol building, Morse transmitted the message, "What hath God wrought?", to Alfred Vail in Baltimore. Just days later, as the Whig party held its national convention in Baltimore, news of the nominations was telegraphed to the Capitol, where politicians learned the results hours before the train carrying the same information could arrive. This stunning demonstration of instantaneous information flow became for America what the dramatic capture of the murderer John Tawell would become for Britain seven months later: a compelling and electrifying display of the telegraph's value.

The simultaneous boom in telegraph technology on both sides of the Atlantic was not merely a coincidence. In 1836, the biggest problem every electrical experimenter faced was suddenly solved. The problem was that a voltaic pile could die in minutes, because hydrogen bubbles formed on the copper plate and smothered the current. The English chemist John Frederic Daniell (1790–1840) cured this by keeping the two metals and their chemical partners separated. He placed a zinc rod in a clear bath of zinc-sulfate, put a copper sheet in a blue bath of copper-sulfate, and let the liquids touch only through a clay wall full of tiny holes. Zinc could dissolve quietly in its own chamber, while copper ions in the outer chamber soaked up the incoming charge and settled onto the plate as fresh metal, leaving no gas to cling and no sudden drop in power. The electric pressure of this type of *Daniell cell* could stay steady at just over one volt for months. The impact of this reliable current source was almost immediate, as a row of such jars would be able to run a telegraph office day and night without attention. Both Cooke and Morse relied upon this innovation in their installations.

Following the success of Morse's 1844 demonstration, the experimental line was placed under the authority of the United States Post Office to be operated as a public service. While a technical marvel, its commercial viability remained an open and critical question. As the government spent its first year grappling with the unfamiliar business of operating a telegraph, Morse and his partners did not wait to see the outcome. In 1845, they formed their own private venture, the Magnetic Telegraph Company. Their strategy was to model the expansion after the railways: building out the "main line" themselves, while licensing patent rights to local entrepreneurs who would build out branches state by state. Not only was their product now proven, but unlike the financial desert Morse had faced earlier, the company was launching into a dramatically more receptive economic climate. The severe credit restrictions imposed by the Bank of England to protect its gold reserves had finally eased, and international capital was once again flowing.

For the Magnetic Telegraph Company, the most important route was the corridor from Washington, D.C. through Baltimore, Delaware, and Philadelphia up to New York City. Ezra Cornell was entrusted as the master contractor for this entire network, subcontracting out sections as needed. In 1840s America, each region came with its own difficulties and restrictions, and rights had to be obtained city by city. Thus, once this first trunk line had been established, in order to expand, it was generally most practical to form a local company within each region to obtain the necessary permits. Each company needed its own capital structure, and so expansion was inherently capital intensive. Cornell began to coordinate with cities, investors, laborers, and suppliers, to set up such companies in order to expand the telegraph network, preferring to be compensated himself in equity in each new venture, in exchange for his efforts.

Using this strategy, Cornell worked his way up the Hudson River to Albany, and then west along the Erie Canal—perhaps the most important commercial artery in the United States at the time. This waterway carried immense quantities of grain, lumber, and goods between the Great Lakes region and the port of New York. A telegraph line along this route gave merchants and financiers in New York City near-instantaneous information about the flow of these vital commodities, allowing them to corner markets and make fortunes.

Cornell built his empire rapidly. He taught his formula to trusted associates, who became key partners in his enterprise. John James Speed, Jr. (1803–1867), once a

Buffalo grain dealer, and Cornell's brother-in-law, helped raise subscription money, underwrote materials and, on routes he knew best, signed as prime contractor himself. Jeptha Wade (1811–1890), once a traveling portrait artist, organized the pole gangs, also taking on the contractor role at times. As soon as Speed reported the next fifty miles of pledges, Wade packed his camp tools and led the crew forward to the fresh survey stakes. Cornell supplied wire, patents, and the legal boilerplate that turned each stretch of right-of-way into a stand-alone corporation. Because state legislatures would grant a franchise only to residents, the trio left behind a mosaic of telegraph company names — Lake Shore, Erie & Michigan, Ohio Indiana & Illinois — all paying the Morse royalty and issuing a block of shares proportional to the risk each one had accepted for the job. In this way, the triumvirate accumulated overlapping stakes mile by mile. By running wires along canal towpaths, stage roads, and even bare survey lines, they frequently gave a town telegraph service weeks before the first rails or locomotives arrived.

Francis Smith, despite being a co-owner of the Morse patents, tried to outflank his partners in New England by stringing a Boston-to-Portland line equipped with an alternative chemical printing method. At the same time Henry O'Reilly (1806–1886), initially a licensed builder, not only sprinted west from Philadelphia into the Ohio and Mississippi valleys but also doubled back along the main Washington-to-Boston corridor, nailing a second set of cross-arms to the existing poles and marketing the circuits as "independent telegraph roads". Morse's Magnetic Telegraph Company sued to protect his monopoly. In a final appeal, O'Reilly v. Morse (1854), the United States Supreme Court upheld the broad claim that any single-wire pulse system was covered by Morse's patent, halting O'Reilly's extensions and warning investors off alternative schemes, including Smith's. By 1855 the remains of these companies were sold to two fully licensed systems: the burgeoning American Telegraph Company in the northeast, and Southern Telegraph Company in the southeast.

Well before the Supreme Court confirmed Morse's monopoly, Cornell and his partners already knew that their stack of small charters was limiting their access to capital. In 1851, Ezra Cornell needed a single charter that could legally own telegraph wire in both Illinois and Missouri, to satisfy railroad demands, so that he could finance a line from Chicago to Saint Louis. Rochester banker Hiram Sibley (1807–1888) already possessed a small, fully paid-in New York corporation — the Rochester Telegraph Company. The Rochester Telegraph Company was a short, local carrier — twenty-two miles of wire chartered in 1849 to link Rochester's flour mills with Lake Ontario shipping points — whose charter allowed amendments. Sibley and Cornell rewrote the charter, increased its authorized capital to $750,000, and re-filed it in Albany under a new name: New York & Mississippi Valley Printing Telegraph Company. Recent state statutes had begun to recognize "foreign" corporations if those firms carried substantial paid-in capital, so the idea of a multi-state owner was now legally possible, and the large share authorization in the charter persuaded county boards on both sides of the Mississippi to grant pole permits to one company, instead of two. Cornell could sell New York & Mississippi Valley shares in each depot town along the projected Chicago & Alton rail grade. Those local subscriptions, together with short-term bank notes he secured by pledging stock from his earlier companies, financed the wire that finally reached Saint Louis on April 1, 1854.

With the Supreme Court's O'Reilly v. Morse decision handed down two months earlier, the legal haze that had kept financiers wary vanished. Sibley returned with a bigger proposal: fold Cornell's Lake Shore, Speed's Illinois & Mississippi, Wade's Ohio, Indiana & Illinois, and Sibley's own Rochester line into the New York & Mississippi Valley charter and relist the enlarged stock on Wall Street. Cornell agreed because consolidation would convert his scattered, thinly traded certificates into a single security that banks would accept at par. Cornell was also content to be the company's largest shareholder, wielding his influence on the board, while letting Sibley, as president, handle the politics necessary to build their business. On November 9, 1855, Cornell, Speed, Wade, and Sibley completed a share-for-share exchange. New York & Mississippi Valley's capital swelled, and five months later the board adopted the name Western Union. From that point forward Cornell could finance new construction by pledging his Western Union shares, building new mileage out of those proceeds, and selling the finished project back to the parent company for more stock — a partnership that suited both builder and banker.

Almost from the moment gold was discovered in California in 1848, Americans spoke of a wire that would bind the states with "electric light and iron wire". The Pony Express would soon promise ten-day letters — but merchants, newspaper editors, stockjobbers, and government officials all wanted minutes, not weeks. Every western crisis underscored the gap: the 1850 California statehood debate, the 1851 Apache raids, the 1857 Utah war. Telegraph promoters from Saint Louis to San Francisco floated schemes — some serious, most fantastical — to leap the "Plains" and the "Sierra", yet those with the capital, including Cornell, balked: eleven hundred miles of unpoliced desert (plus another seven hundred miles of frontier) was an expensive place to hang hundreds of tons of iron wire. By 1858, the United States Congress was circulating draft subsidy bills and hearing testimony on how a transcontinental telegraph might be financed.

In mid-1858, Western Union's president, Hiram Sibley, dispatched Edward Creighton (1820–1874) to California. There, Creighton met with James Gamble (1826–1905), chief engineer for most of the telegraph wires that had yet been built in California. Gamble swore he could push a wire east from Carson City as fast as any crew could push west. Darius Ogden Mills (1825–1910), perhaps California's wealthiest resident, committed to front the cash himself for this western push. Creighton and Gamble could meet in Salt Lake City, where Brigham Young (1801–1877) agreed to host the splice and protect the line, in return for a major telegraph station of his own, as well as for the guaranteed economic opportunity of providing the labor and poles. Creighton's report, together with the promise of government subsidies, persuaded Cornell to risk his Western Union stock as collateral for supplies. Within weeks, over three thousand coils of iron "No. 9" wire and over sixty thousand glass insulators were on order — a little over half billed to Omaha, and the rest to Sacramento, where parallel local charter companies were established. Congress, seeing the private capital already in motion, and sensing the looming need for wartime control of the frontier, passed the Pacific Telegraph Act in June 1860, promising free right-of-way and a ten-year message contract (up to $40,000 per year) to the first company that finished.

As the poles advanced, Creighton and Gamble exchanged weekly progress updates via Brigham Young's clerk in Salt Lake City. Creighton's wire touched Salt Lake just after dawn on October 18, 1861. Six days later, Gamble's line rolled in from the west. The last splice was made in the attic of Young's "tithing-house" at two o'clock in the afternoon. Congratulations were flashed up and down the wire. The Pony Express shut down two days later. The telegraph network became the nerve of federal command almost overnight. Within weeks, military traffic — suddenly urgent in the first year of the American Civil War — was moving on the private line Cornell and Mills had underwritten. The United States government had asked for a coast-to-coast wire and allowed ten years; it was delivered in sixteen months, six of them under wartime conditions.

The completion of the Pacific line in 1861 placed Western Union on a path toward domination. After the war, only one major obstacle prevented a true national monopoly: the powerful American Telegraph Company, which controlled the most lucrative east coast networks. In an audacious final campaign in 1866, Western Union absorbed this last great rival, along with the rest of its competitors, in a series of massive stock-swap deals. With this "great consolidation", the telegraph wars were over in the United States, and Western Union had achieved the continent-spanning monopoly that its proprietors had long envisioned.

While Western Union had been focused on connecting the east coast to the west coast, American Telegraph Company, run by engineer Marshall Lefferts (1821–1876), had controlled the dense Eastern seaboard, and saw the ultimate prize as connecting the New World to the Old World. Lefferts's focus was international, forming a strategic alliance with a company founded for the sole purpose of laying an undersea trans-Atlantic cable: the Atlantic Telegraph Company. This company was the passion project of an obsessive entrepreneur named Cyrus Field (1819-1892), who had already made a modest fortune manufacturing paper. Field raised immense capital from American and British investors, relentlessly pushing the project forward despite repeated, catastrophic failures. The deal was that Fields's company would own the cable itself, while Lefferts's company would be granted the exclusive rights to connect to the cable and handle all of its traffic in the United States.

In January of 1854, Cyrus Field sailed north to inspect the skeleton of a half-finished Newfoundland telegraph line. For the trip, he had purchased a copy of Matthew Fontaine Maury's (1806–1873) *A Chart of the North Atlantic Ocean* (1854), which plotted thousands of United States Naval measurements and revealed a shallow, gently rising ridge, running almost dead-straight from Newfoundland to the southwest coast of Ireland. If a cable were laid on that Telegraphic Plateau, it would settle under its own weight into a soft bottom in which "gravity alone", Field wrote, "would cradle it beyond the reach of chafe or tide". Field bought the charter for the bankrupted telegraph, convinced of the vision he had seen. He sailed at once for London, where the Newfoundland creditors passed him to Britain's foremost submarine telegraph promoter, John Watkins Brett (1805–1863), famous for successfully connecting England to France by wire, under the Strait of Dover, in 1851. Brett presented Field to undersea cable engineer Charles Tilston Bright (1832–1888), who was an acquaintance of physicist George Gabriel Stokes through the British Association for the Advancement of Science. Bright escorted his new American client to a Michael Faraday lecture at the Royal Institution, where Stokes introduced Field to William Thomson. Field managed to enlist Brett, Bright and Thomson as technical guarantors, and, fortified by their calculations, secured the loan of HMS Agamemnon from the British Admiralty and a matching commitment of USS Niagara and traffic payments from the United States Congress by the end of 1857.

Field made his proposal to the Admiralty in late November 1856, and a Treasury minute approving the naval ship followed on February 2, 1857. Britain's endorsement came startlingly fast because the country had just suffered a costly, well-documented failure. During the "Crimea campaign" London first ordered, then tried to cancel, a winter assault on the Russian Baltic fortress of Kronstadt. The reversal, dispatched by steamer on December 15, 1854, arrived almost four weeks later, after the fleet had already advanced and ice had blocked the channel. Coal, ammunition and an expeditionary squadron spent the season in frustration. The press publicly criticized the waste, and an internal report called the delay "a grave hazard to the Baltic Expedition". Determined not to repeat the mistake, Royal Engineers that spring reeled out a three hundred forty mile insulated overland wire from Balaklava siege lines to the telegraph office at Varna, tying the Crimean front into Europe's network, and cutting message time to London from days to hours. The contrast was explicit in the 1856 military post-mortem: telegraphy could save a campaign, but an ocean gap could lose one. When Field, Bright and Thomson arrived with a design that promised to close the Atlantic gap, Admiralty officials saw a solution to prevent another fiasco.

The project was bolstered by the knowledge that a cable landing in Ireland would connect seamlessly with a European continent that was, ironically, standardizing on the Morse system. To solve the chaos of its patchwork of incompatible national systems, the powerful Austro-German Telegraph Union had adopted a version of Morse for interoperability in 1851, creating a domino effect that established a common language for international traffic. Even Britain had already equipped its Dover station with Morse keys so cross-Channel messages could be sent to French Calais without needing to be rekeyed.

The first voyage left Ireland in the summer of 1857 with USS Niagara and HMS Agamemnon, each carrying more than half of a cable whose iron armor wires were wound very tightly around the insulated copper core. The plan was that the two ships would meet at a designated rendezvous point, splice their respective ends of the cable together, and then sail in opposite directions – the Agamemnon eastward towards Valentia Bay in Ireland, and the Niagara westward towards Trinity Bay in Newfoundland. The first attempt failed within minutes, as they were lowering the spliced cable into the sea. On the next two attempts, after laying cable for days, a heavy swell made the ships roll differently; their brakes no longer paid the cable out at the same speed and the stiff line snapped. Without enough wire to continue, the engineers recorded three lessons for next time: wind the armor with a more flexible "soft-lay" spiral, add brakes that would keep steady tension, and, if ever a ship large enough could be found, carry the whole cable in one hull.

As the team prepared for another voyage, William Thomson took a more active role in tackling some of the challenges. While Gustav Kirchhoff's 1857 telegrapher's equations had brilliantly described the wave's theoretical speed in an ideal system, Thomson's earlier work had addressed *signal retardation* — the way a whole voltage pulse would be carried through a circuit. He showed how a submarine cable would suffer from two compounding issues: immense electrical resistance, and massive capacitance. Both increased with distance, so their combined effect meant a cable twice as long was four times as retarded, smearing any sharp pulse into a faint, delayed swell of current. To detect the "ghost" of a signal that his theories predicted, Thomson built upon the optical lever, developed by Johann Christian Poggendorff (1796–1877) and used by Carl Friedrich Gauss and Wilhelm Weber in their 1833 telegraph. He replaced Gauss's heavy magnet with a tiny mirror assembly, and instead of viewing a scale through a telescope, he projected a focused beam of light off the mirror onto a distant scale. The tiniest twist of the magnet moved the spot of light several inches, creating a sensitive *mirror galvanometer*.

In 1858, they tried again with softer cable and self-adjusting steam brakes, though still with the same two ships. Before setting sail, William Thomson equipped both ships with his refined mirror galvanometers and drafted an elaborate continuity drill. Every fifteen minutes one ship switched its battery onto the wire and tapped the two-letter signal "GA" (for "go ahead"). The other ship, with its battery disconnected, watched for the returning flash of lamp-light on the mirror scale. If the deflection was weak, the sender added a tray of ten cells (around eleven volts). The first three attempts that summer ended when storms at sea snapped the cable, but on the fourth try both vessels arrived at their shores in unbroken contact. For three weeks after landfall the wire carried traffic, including Queen Victoria's greeting to President Buchanan, across the Atlantic on a whisper of electrical current.

Trouble began almost as soon as the first public messages flew. On August 5, 1858, Agamemnon had landed its end in Valentia Bay, Ireland, and William Thomson — having spent a week verifying the mirror readings — then had the cable connected to a land line to London so he could show the board the bright, steady flashes on his scale. Chief electrician Edward Orange Wildman Whitehouse (1816–1890) stayed behind in the wooden shore hut, convinced that the line could — and must — be driven faster. Wildman was a retired surgeon who, as a small, early shareholder, had managed to get himself appointed as the project's operational authority. He had amateur experience with shorter landlines, where higher voltage was used to overcome resistance. He believed a "brute force" approach was needed and had built massive, five-foot-long induction coils capable of generating up to two thousand volts.

On August 16, 1858, Wildman replaced the Daniell battery with an induction coil that catapulted the working voltage from roughly two hundred volts to more than twelve hundred. The next day, Thomson watched as his London mirror meter sagged by half. He telegraphed Valentia asking what had changed and Whitehouse replied that he had substituted "a more energetic arrangement of cells". Thomson sent a formal letter to the directors, warning that excessive voltage would endanger the insulation. Whitehouse dismissed Thomson's mathematical theories as overly academic, and answered by increasing the voltage further. During the last week of August the signal dwindled day by day. Thomson rushed back to Valentia, arriving to find Whitehouse still insisting that the cure was stronger current. Early on September 2, the mirror beam died altogether, turning the world's most expensive conductor into a useless heater. The board was finally convinced that Thomson's conclusion was correct: increasing the voltage makes noise and insulation damage increase faster than the signal. Whitehouse was dismissed within weeks of the incident, and Thomson was given full authority over the project from that point forward.

After the glorious but brief success of the 1858 cable, the project was left financially ruined and stalled for years by the immense cost, exacerbated by the outbreak of the American Civil War. The venture was kept alive through the constant efforts of its founder, Cyrus Field, and its key British director, the industrialist John Pender (1816–1896). In 1864, Pender rescued the project with an ingenious plan. First, he combined the two main cable-making factories in England into one giant, reliable company called Telcon. With the manufacturer securely in his pocket, Pender was then able to form another operating venture, the Anglo-American Telegraph Company, to raise the capital and charter the only ship big enough for the job: Isambard Kingdom Brunel's twenty two thousand ton SS Great Eastern. Previously unfinished and tied to a failed passenger-line scheme during the earlier attempts, the scheme commercially collapsed in 1863, and the ship became available at a bargain. In 1865, they set sail again, but more than halfway through, the cable snapped and was lost at the bottom of the sea. Because Telcon's sale of the first cable had been profitable, the overall venture had enough capital to try again. The next summer, in 1866, the Great Eastern flawlessly laid a wire from Ireland to Newfoundland. In an incredible display of skill, the ship then returned to the middle of the ocean, grappled the cable they had lost the year before from two miles deep, spliced on fresh cable, and completed the line to shore. By September 1866, two working cables were connecting the continents.

By the late 1860s, Britain's telegraph scene was a maze of individually chartered companies, whose ownership rolled up to a small, powerful cartel, much as had happened in America. However, unlike in America, where Congress's experimental line of 1844 was returned to private hands after the Post Office failed to operate it profitably, the British Parliament decided that the system would better serve the public as nationalized infrastructure. The Telegraph Act of 1868 authorized a government buy-out, and an amendment the following year set a single nationwide price, effectively folding all domestic lines into a single state network run by the General Post Office. Despite nationalization, the battle for international traffic remained private and competitive, where a fractional gain in signalling speed could decide who carried the lucrative press dispatches.

It was into this landscape that eighteen-year-old Oliver Heaviside entered the telegraph industry in 1868 as a junior operator for the Anglo-Danish Telegraph Company — a Newcastle-based venture whose North Sea cable to Denmark suffered from the same blurred pulses that had plagued the trans-Atlantic submarine line. In 1869, three Scandinavian cable firms, including the Anglo-Danish, merged to form the Great Northern Telegraph Company, in order to pool capital to build massive wires from Russia to Asia. After the merger, Heaviside — whose family lived in northeast England — was transferred to nearby Newcastle-upon-Tyne, and promoted to senior operator. By late 1871, Heaviside had become the Chief Night Operator of the only British landing station for the cable through Denmark to Russia.

In 1872, Heaviside began writing short papers addressing practical problems he encountered in daily telegraph operations. His first published piece, *Comparing Electromotive Forces* (1872), appeared in the "English Mechanic", and demonstrated how to quickly determine which of two batteries was stronger by inserting a variable resistor and galvanometer between them, then adjusting the resistor until the galvanometer showed no deflection. Later that same year, he published another piece in the English Mechanic, proposing an improved method for configuring his uncle's well-known measuring device, the *Wheatstone bridge*. He refined this work into a more rigorous mathematical paper which appeared in the prestigious Philosophical Magazine in February 1873. This paper caught the attention of William Thomson, who read it on a train journey and was sufficiently impressed that, on his next trip through Newcastle, he made a point to visit Heaviside personally. Thomson encouraged Heaviside to continue writing, and suggested he apply for membership to the Society of Telegraph Engineers. Buoyed by such praise, Heaviside applied, but the council initially rejected his application, privately noting that they "didn't want telegraph clerks". Stung by the rejection, Heaviside appealed directly to Thomson, who then sponsored a renewed application. Early in 1874, the council reversed its decision, electing the twenty-three year-old Heaviside as an Associate Member — a rare honor for someone without university training.

Heaviside first encountered James Clerk Maxwell's *Treatise on Electricity and Magnetism* in the Newcastle Literary and Philosophical Society during the summer of 1873, only a few months after it had been published. That autumn, Maxwell himself printed an "Errata and Addenda" sheet, containing verbatim three formulas from Heaviside's Wheatstone bridge paper, crediting him by name. In May of 1874, with night duty worsening both Heaviside's hearing and his nerves, the twenty four year old operator found that both the mathematician he most admired and the engineer he most respected had publicly stamped his first mathematical paper. The combination of his physical condition and his desire to devote himself to full-time study of the "heaven-sent" Maxwell (as Heaviside later wrote), led him to resign from his work in the telegraph office on May 31 and move back into his parents' Camden Town home.

Heaviside's first major target was achieving reliable *duplex* operation — simultaneous two-way communication on a single wire. While others, like the American inventor Joseph Barker Stearns (1831–1895), had achieved rudimentary duplex, their systems often failed on long submarine cables or at higher signaling speeds. The core problem was that an operator's powerful outgoing transmission would overwhelm their own local receiver. The existing solution involved a balanced bridge circuit where the outgoing current was split: half went down the real cable, and the other half went into a *dummy circuit* (an artificial line) calibrated to match the cable. However, these early dummy circuits only accounted for resistance and capacitance. Heaviside realized that as speeds increased, the cable's inductance — the electromagnetic inertia that resists changes in current — became significant. In his studies *Duplex Telegraphy* (1873 & 1875), he demonstrated that the dummy circuit must mirror the full impedance of the cable, including inductance. By precisely balancing all these factors, the local receiver remained deaf to the outgoing transmission but perfectly sensitive to incoming signals. By 1878, field trials were successfully following his prescriptions.

Living rent-free and sustaining himself on meager fees from articles in trade journals like The English Mechanic and The Telegraphic Journal, Heaviside dedicated himself to mastering Maxwell's dense theories. He relentlessly applied this emerging field view to the practical realities of the wire. In *On the Extra Current* (1876), he revisited the established *telegrapher's equations* developed by Gustav Kirchhoff and William Thomson. Previous models had focused primarily on capacitance (C) and resistance (R), recognizing that a submarine cable with its thin insulation and the sea water's conductivity behaved itself like an enormous capacitor. Kirchhoff had included inductance (L) in his model, whereas Thomson had ignored it, and neither accounted for the "leaky" conductance (G) through the insulation. By incorporating all four elements — R, L, C, and G — Heaviside gained the first complete mathematical picture of how signals propagated, distorted, and reflected along the line. He pressed further in *On the Speed of Signalling through Heterogeneous Telegraph Circuits* (1877), providing formulas for lines whose properties varied from section to section — the exact situation faced by real-world inter-city networks.

Throughout the early 1880s, Heaviside was engaged in a monumental act of intellectual synthesis. Finding the established mathematical formalisms to be physically unintuitive, he hacked away at the scaffolding of Maxwell's work to reveal the underlying physical reality. This effort culminated in the elegant *vector calculus* that replaced the cumbersome older methods. As Heaviside published his findings in a standing column in The Electrician, his work polarized the engineering community. The "practical men", epitomized by William Preece (1834–1913), the chief engineer of the Post Office, dismissed Heaviside's theories — especially the importance of inductance. However, a small group of academic physicists, the Maxwellians, recognized the profound implications of his work. George Francis FitzGerald (1851–1901) of Trinity College Dublin, and Oliver Lodge (1851–1940), a prominent physicist at the University of Liverpool, became his staunch advocates. FitzGerald and Lodge initiated a correspondence in the mid-1880s, championing Heaviside to the Royal Society and engaging him in vigorous debate, forming a crucial, albeit remote, intellectual community for the isolated theorist.

It was within this period of intense theoretical refinement, formalized around 1887, that Heaviside derived the precise mathematical condition for perfect signal propagation. Distortion occurs because the different frequencies that make up a signal travel at different speeds through an unbalanced cable, "smearing" the pulse. Heaviside proved that if the ratio of the resistance to the inductance precisely equaled the ratio of the leakage conductance to the capacitance ($\frac{R}{L} = \frac{G}{C}$), all frequencies would travel at the same speed, and the signal would arrive perfectly intact. This revelation became critical with the invention of the telephone in 1876; while distorted telegraph clicks were manageable, the complex waveforms of human speech were incomprehensible over long distances. The solution was counterintuitive: the only practical way to restore balance was to drastically increase the cable's inductance. While manufacturing a cable with uniformly high inductance was impractical, discussions among the Maxwellians — notably FitzGerald and Lodge — suggested that Heaviside's condition could be approximated by inserting *loading coils* at periodic intervals along the wire. Heaviside provided the mathematical proof that this pragmatic engineering solution would work, leaving it to others in industry to patent the technology that would make long-distance telephony possible.

Meanwhile, the electrical world was exploding beyond telegraphy, into telephony, and power. The Bell Telephone Company, founded in Boston in 1877 by the Scottish-born Alexander Graham Bell (1847–1922) and his financial backers, quickly wired urban centers. However, engineers faced immediate crises. Telephone signals were highly susceptible to distortion and crosstalk, where the changing electrical field of one wire induced unintended currents in others. The urgency intensified dramatically when Thomas Edison (1847–1931), operating out of his Menlo Park, New Jersey laboratory, demonstrated his durable carbon-filament lamp in October 1879. Simultaneously, firms like Germany's Siemens & Halske, founded by Werner von Siemens (1816–1892), were pushing high-current systems for urban arc lighting. These new demands — high frequencies in telephony and high power in lighting — required exactly the kind of sophisticated analysis of inductance and capacitance that Oliver Heaviside was developing.

The demand for industrial power ignited the War of the Currents, a conflict that hinged on a fundamental mathematical truth about electrical transmission. Edison championed *direct current* (DC), but his system faced the tyranny of resistance. Wires inevitably lose energy as heat, and this loss is calculated as the current squared times resistance (I^2R). Edison's DC system operated at relatively low voltages (100–220V). To transmit significant power at low voltage, the current must be extremely high. This high current resulted in enormous energy losses over distance, demanding impractically thick copper cables or a power station virtually every mile.

The revolution came with *alternating current* (AC), championed by the visionary Serbian-American engineer Nikola Tesla (1856–1943) and commercialized by George Westinghouse (1846–1914), the Pittsburgh industrialist known for the railway air brake. AC offered a decisive advantage: the *transformer*. Transformers could efficiently "step up" voltage for transmission and "step it down" for safe use. By drastically increasing the voltage, the current is proportionally decreased. Because power loss is based on the square of the current, the efficiency gain was enormous: increasing voltage a hundredfold meant one ten-thousandth the loss.

By the early 1890s, the intense competition, wasteful infrastructure duplication, and staggering costs of patent litigation proved unsustainable. The landscape was dominated by Edison General Electric in New York and the highly profitable, AC-focused Thomson-Houston Electric Company of Lynn, Massachusetts. This chaotic rivalry was anathema to James Pierpont Morgan (1837–1913), the dominant Wall Street financier backing Edison. Morgan despised disorganized competition and sought to impose order through consolidation. In 1892, he orchestrated a merger, creating the monolithic General Electric. Thomson-Houston, with its superior AC patents and stronger financials, took control under the leadership of the shrewd former shoe manufacturer, Charles Albert Coffin (1844–1926). In a stark illustration of capital prioritizing profitability over personality, the "Edison" name was dropped from the masthead. Edison's stubborn, public crusade against AC — denouncing it as dangerously unsafe, sponsoring gruesome animal electrocution demonstrations, and lobbying municipalities to ban it — was now viewed by the Morgan-backed board as a liability.

The definitive verdict in the War of the Currents was delivered at Niagara Falls. In 1891, the Cataract Construction Company, formed to develop hydroelectric power at the falls, convened the International Niagara Commission, a blue-ribbon panel of the world's leading scientists, to judge proposals from across Europe and America. The commission was led by William Thomson, the towering British physicist who, ironically, had initially favored DC. They reviewed designs ranging from compressed air systems to Edison's massive DC scheme. But the physics of resistive loss proportional to current-squared was undeniable. Westinghouse, armed with Tesla's patents, proposed a polyphase AC system whose transmission efficiency dwarfed all rivals. The commission recommended AC in May 1893. Following the spectacular AC-powered illumination of the Chicago World's Fair later that year, Westinghouse secured the contract. On November 16, 1896, the switches were closed, and power surged from Niagara to Buffalo (twenty-two miles away), heralding the triumph of the AC grid.

While the battles over electrical infrastructure raged on land, the theoretical foundation laid by James Clerk Maxwell was preparing the way for a far more radical technology: the control of the airwaves. Maxwell had explicitly predicted that electromagnetic energy would propagate through empty space as waves traveling at the speed of light, but for decades his theory remained obscured by dense mathematics. In 1887 the German physicist Heinrich Hertz, working directly from Maxwell's ideas, built a simple spark-gap apparatus that generated and detected such waves, proving their existence. Meanwhile, Oliver Heaviside's diligent effort to synthesize Maxwell's work provided necessary clarity. Hertz's demonstration showed that the waves were real, while Heaviside's reformulation made the theory broadly intelligible, ensuring that the significance of Hertz's discovery could be grasped and ultimately harnessed.

The implications of these experiments were immediately understood across the globe: information could be transmitted without wires. This ignited a fervor of experimentation, but this frontier was also a direct threat to the existing global order. In the late nineteenth century, British power was projected through a submarine lattice of nearly one-hundred-fifty-thousand kilometers of copper cable, controlled largely by the Eastern & Associated Telegraph Companies. Ninety percent of the Empire's diplomatic and commercial traffic flowed through London. While figures like the British physicist Oliver Lodge (1851–1940) and Nikola Tesla explored the potential of wave propagation, it was the young Italian-Irish inventor Guglielmo Marconi (1874–1937) who relentlessly pursued the goal of bypassing this monopoly entirely.

The pivotal moment arrived on December 12, 1901, when Marconi, stationed in Newfoundland, detected a faint signal transmitted from Cornwall, England. This first trans-Atlantic radio signal — achieved just thirty-five years after the first reliable cable — proved that the curvature of the Earth would not limit wireless communication, instantly transforming it from a possibility into a strategic asset. This achievement stunned the scientific community, not just because of the distance, but because it defied expectations. Electromagnetic waves, like light, should travel in straight lines, shooting off into space rather than hugging the curvature of the Earth.

In 1902, both Heaviside and the American electrical engineer Arthur Kennelly (1861–1939) independently proposed a solution: there must be a conductive layer in the upper atmosphere acting as a mirror, reflecting the radio waves back down to Earth, allowing them to propagate over the horizon. The idea was ingenious but unproven. No one at the time had any means to directly probe the upper atmosphere, so the "reflecting layer" remained a speculative construct rather than an observable fact, and many scientists were reluctant to accept it until experimental confirmation came decades later.

At the same time, the speed of light itself had become a pressing puzzle. Maxwell's equations implied that electromagnetic waves always moved at a fixed rate through space, but by analogy with sound or water, one would expect that an observer chasing or running away from a wave should measure a different speed. In 1887, Albert Michelson (1852–1931) and Edward Morley (1838–1923) built an apparatus to measure the apparent difference. A single light beam was divided into two paths set at right angles, reflected back, and recombined. With the Earth's presumed motion through the ether, they expected the two returning beams should have fallen out of step. Instead they remained aligned. The anticipated shift never appeared, leaving them to conclude the observed speed of light was unchanged in every direction.

In 1888, Heaviside, in the midst of reworking Maxwell's system into vector form and analyzing the fields of moving charges, discovered that the electric field around a charge in uniform motion is squeezed along its line of travel, spreading sideways into a flattened ellipsoid shape. His aim was to understand how fields behave around charges in motion, but the result also provided the sort of distortion effect that, if applied to the forces holding matter together, could make Michelson and Morley's null result make sense.

In early 1889, George FitzGerald wrote to Heaviside about this idea, and in May of that year he proposed that whole moving bodies would contract in their line of motion by precisely the required amount. In 1892, Hendrik Lorentz (1853–1928) reached the same conclusion, but cast in such a way that the transformation applied not only to the forces within matter, but to the spatial coordinates themselves.

In 1905, Albert Einstein (1879–1955) took a different approach to resolving Michelson and Morley's null result by simply treating it as a postulate: every observer must measure the same constant speed of light. He coupled this with what Henri Poincaré (1854–1912) called the *principle of relativity* — that the laws of physics take the same form for all observers in uniform motion. Rather than explaining contractions in the ether, Einstein discarded the idea of such a medium as unnecessary — an unsettling move to some. Superfluous or not, however, control over the ether was still hotly contested.

Guglielmo Marconi's transatlantic signal in 1901 was far more than a technological triumph; it was a geopolitical earthquake that threatened Britain's global communications monopoly. This resonated deeply within the German Empire, which, under Kaiser Wilhelm II (1859–1941), was pursuing an assertive foreign policy known as *weltpolitik* (world politics). Germany, a rising industrial power unified in 1871, sought its "place in the sun" by expanding its navy and securing overseas colonies to rival the British Empire. Communications independence was vital to this ambition. Recognizing that reliance on British cables meant Germany could be isolated in an instant, the Kaiser pressured Germany's rival electrical giants, Siemens & Halske and Allgemeine Elektricitäts-Gesellschaft (AEG) to cooperate, and in 1903 they formed the joint venture Telefunken to challenge the Marconi Company's dominance.

For Britain, the danger was that wireless signals could leap across oceans without touching its undersea cables, stripping London of its ability to monitor, tax, or cut off messages at will. The government's answer was the Wireless Telegraphy Act of 1904, which required a license for every transmitter and brought the airwaves under state authority — a legislative enclosure designed to preserve the same strategic control Britain had long exercised through its cable empire. Germany retaliated with more engineering; in 1906, Telefunken opened the powerful Nauen Transmitter Station.

The next decade saw the airwaves descend into chaos, jammed with competing signals from commercial operators, navies, and enthusiastic amateurs. Early attempts at order, such as the International Radiotelegraph Conference of 1906 in Berlin, standardized the "SOS" distress call and set basic operating rules, but comprehensive frequency coordination remained elusive until after the Titanic disaster in April 1912. While the tragedy primarily stemmed from operational failures — most critically, the nearest ship's radio operator being off-duty — corporate and military lobbyists seized the opportunity, promoting a narrative that "chaotic" amateur interference had hindered rescue efforts. This provided a ready pretext for governments to clamp down. The United States Radio Act of 1912 mandated federal licensing for all transmitters, making it illegal to broadcast without permission and effectively sidelining independent operators in favor of centralized control.

Meanwhile, the wireless arms race escalated. By 1913, Telefunken had upgraded Nauen to a massive one hundred kilowatts and established a chain of stations stretching through Africa, notably the high-power Kamina station in Togoland. Germany had achieved a functional, independent global ether route. Britain's answer, the proposed Imperial Wireless Chain to knit the Empire together, became entangled in politics. In 1912 the government hastily awarded the lucrative contract to the Marconi Company, and it soon emerged that senior ministers, including the Attorney General and the Chancellor of the Exchequer (and future Prime Minister) David Lloyd George (1863–1945), had bought shares in American Marconi just before the public announcement. The resulting Marconi Scandal bogged the project down, leaving Britain heavily reliant on vulnerable submarine cables as war loomed in 1914.

The first blows of The Great War in August 1914 were aimed not at armies, but at communications infrastructure. The British cable ship Alert immediately severed Germany's five Atlantic cables and six North Sea cables. This was strategically crucial despite Germany's advanced wireless capability. Early longwave wireless, though powerful, had limited bandwidth, was prone to atmospheric disturbance, and was inherently insecure — a broadcast medium, easily intercepted. Cables remained essential for reliable, high-volume, and sensitive traffic. Britain's strategy was to deny Germany secure communication and force them onto the airwaves or onto neutral countries' cables, both of which the Admiralty was actively monitoring. At the same time, British forces moved against Germany's wireless network; the crucial Kamina station was neutralized by the end of August 1914 and taken out of service.

This strategy of communication isolation yielded the greatest intelligence coup of the war. In January 1917, German Foreign Secretary Arthur Zimmermann (1864–1940) needed to send a sensitive proposal to Mexico, suggesting an alliance against the United States. Germany's own wireless network was heavily monitored and unreliable, and its direct cables were cut. The most secure route available ran through the United States, which — still neutral — allowed Germany to send encrypted diplomatic traffic via State Department facilities. Those cables were physically routed through Britain. Germany assumed diplomatic courtesy guaranteed privacy, but British intelligence (Room 40) was already tapping the lines. The Zimmermann Telegram was intercepted and deciphered; its publication inflamed American opinion and helped propel the United States into the war in April 1917.

The war solidified the electromagnetic spectrum as vital strategic infrastructure, transforming it from a chaotic commons into the property of the state and its corporate partners. Upon America's entry in 1917, the Navy seized control of all radio operations, shutting down independent experimenters and forcing industrial players to pool their patents for wartime production. This climate of forced centralization also provided the pretext for eliminating potential disruptions. Wardenclyffe — Nikola Tesla's Long Island laboratory and nearly two-hundred-foot tower, designed both for transoceanic wireless communication and, in his own vision, for the wireless transmission of power — had already been crippled when James Pierpont Morgan withdrew further financing in 1903. Morgan and other financiers regarded Tesla's plan to broadcast energy without meters as a direct threat to the emerging monopolies in communication and electricity. By 1917 the tower stood unfinished, idle, and debt-burdened. Its demolition that year was publicly justified on national-security grounds, but in practice it also liquidated assets and conveniently erased a project that challenged centralized models of control.

When the war ended, the United States Navy declined to relinquish control of the airwaves, deeming radio too critical to be dominated by the British-controlled Marconi company. Under government pressure, General Electric bought out Marconi's American assets. To end paralyzing patent disputes and consolidate the technology, General Electric, Westinghouse, the American Telephone and Telegraph

Company (AT&T, founded in 1885 to build a national long-distance network and monopoly successor to the original Bell company), and the United Fruit Company (which maintained one of the largest private radio networks in the world to coordinate its Caribbean shipping fleet) pooled their key patents. In November 1919 they formed the Radio Corporation of America (RCA) — a state-sanctioned cartel designed to dominate the radio age.

The consolidation of infrastructure after the war led directly to the consolidation of knowledge. The era of open scientific inquiry gave way to the industrial laboratory — Bell Telephone Laboratories (Bell Labs), formally created in 1925 as the research arm of AT&T, and later RCA's own research division. Their purpose was not the open exchange of ideas but proprietary advantage. Results were fiercely guarded as trade secrets; complexity itself became a barrier to entry, ensuring that only specialists within the corporate structure could innovate. Independent inventors were often crushed, as illustrated by the decades-long patent battle RCA waged against the television pioneer Philo Farnsworth (1906–1971). There was little incentive to make the technology transparent. Devices were built as "black boxes" that seemed to work like magic, cultivating a vast population of passive consumers rather than active creators. The Bell System exemplified this philosophy of systematic obfuscation, keeping technical manuals strictly internal until compelled by an antitrust suit in 1956. Combined with the inherent abstraction of invisible fields of force — expressed in the demanding language of calculus, too often taught as a patchwork of rules rather than a coherent grammar — this deepened a cultural sense of intimidation before complex technology.

The priority of control over understanding, and profit over credit, is starkly illustrated by the fate of Oliver Heaviside. His theoretical derivation of the *distortionless condition* (1887) had held the key to long-distance telephony, essential for AT&T to extend its monopoly across the United States. But Heaviside never patented the idea. George Campbell (1870–1954), an engineer at AT&T, worked out a practical method using "loading coils" to add inductance to the line in 1899, while the Serbian-American physicist Mihajlo Pupin (1858–1935) independently patented the approach in 1900. AT&T, intent on securing its monopoly over the technology that made its long-lines network viable, bought Pupin's patent for hundreds of thousands of dollars.

While vast fortunes were built upon his mathematical frameworks, Heaviside lived and died in relative poverty. A reclusive figure who cared little for money and often feuded with the engineering establishment, he prized intellectual recognition above all else. When AT&T offered him a token payment for his role in the loading coil, he refused, remarking that giving him the credit he deserved would "interfere…with the flow of dollars in the proper direction". In the 1890s his allies, including George Francis FitzGerald, William Thomson, and Oliver Lodge, successfully petitioned the British government to grant him a small civil pension. Despite his isolation, the truths he uncovered could not be contained. In 1924, just a year before his death, the British physicist Edward Victor Appleton (1892–1965) conducted experiments that confirmed the existence of the reflective atmospheric layer that Heaviside and Arthur Kennelly had predicted decades earlier — the *ionosphere* — which made global radio communication possible.

And so goes the story of how the language of calculus, refined century by century, grew powerful enough to articulate the rich dynamics of electromagnetism. In this language, a telegraph signal may be modeled as an infinite set of tiny voltage changes across each infinitesimal slice of a transmission medium. The received message is then the accumulation of those tiny changes — an expression of the fundamental theorem of calculus. The electromagnetic waves, propagating through a cable, and throughout all of space, are described as a perfect balance between rates of change in electric and magnetic energy — a differential equation. Generation by generation, engineers iterated upon the models, until no fault could be found in their ability to predict novel, real world implications. Such reliable models could be reasoned about on paper, unleashing an explosion of invention and innovation. Inevitably, the quest for understanding gave way to the demand for power and control. Spectrum, voltage, and frequency were parceled out by patent offices and ministries. By the twentieth century, it had become practically illegal to experiment on such topics without permission. The knowledge, once acquired, was actively obscured. Its proliferation did not serve the interests of those who already possessed it. No wonder then, that so many people feel intimidated by these topics today. Yet this intimidation is manufactured. These techniques are the rightful inheritance of humanity — derived from the unassailable truths of nature, and waiting to be taken up again.

Afterword

The remarkable internal consistency and "accountability" one observes in the most successful physical laws — where local actions precisely sum to global effects, as elegantly captured by mathematical frameworks like the generalized theorems of calculus in field theories — may be understood not solely as a direct reflection of an unmediated, objective external reality, but significantly as a product of the inherent structuring principles of human cognition and the methodologies of scientific model-building. The coherent, predictable, and accountable nature of scientific models is, in part, a necessary outcome of the minds' demand for such order. Humans seek to build conceptual frameworks where the "parts sum to a whole", and this drive for coherence profoundly shapes the laws humans formulate and the mathematical language chosen to express them.

This cognitive insistence on coherence becomes particularly evident when the models face empirical challenges. When a well-established and accountable model appears to falter — when the "parts don't sum to the whole" in light of new data — the scientific impulse is often not to immediately discard the principle of accountability itself, but rather to preserve it by "inventing new parts" or refining existing ones. The postulation of length contraction to maintain an invariant speed of light, or Maxwell's introduction of the displacement current to ensure charge conservation and consistency in electromagnetism, are prime examples of this. These were not ad-hoc fixes but were driven by a deep-seated demand to maintain the integrity and predictive power of an accountable system. The very selection and development of mathematical tools, such as vector calculus and the generalized Stokes's Theorem (which guarantees that local changes within a domain are precisely tallied on its boundary), reflect a preference for, and a reliance upon, structures that inherently enforce this kind of logical and quantitative accountability.

Scientific "discovery", then, is understood as a dynamic interplay: external reality provides empirical data and constraints, while the human mind, with its intrinsic need for coherence and its capacity for abstract representation, actively constructs models that are both empirically adequate and logically satisfying. The "universality" and "accountability" of physical laws are thus a testament not only to any underlying order of the cosmos but also to the fundamental ways human cognition is equipped to perceive, process, and make coherent sense of that order. These laws represent successful, robust interfaces between cognitive architecture and the universe, reflecting a co-created understanding where the demand for a consistent, accountable narrative is as crucial as empirical observation.

While mathematics serves as an extraordinarily powerful language for describing the physical universe, its deep study also offers a unique journey into understanding the mind itself. If the very structure of successful scientific models — their "accountability", their reliance on consistent logical frameworks, and their elegant symmetries — is partly a reflection of the human mind's inherent demand for coherence, then mathematics itself can be seen as a mirror reflecting the fundamental architecture of human cognition. The axioms chosen, the logical rules that are found compelling, the patterns the mind is drawn to discern and formalize, and the very notion of "proof" that satisfies the intellect, all speak volumes about the innate "operating system" through which humans process reality and construct understanding. By engaging deeply with mathematics, one is not just learning about numbers, shapes, and structures, but also conducting an introspective examination of the very tools and frameworks their mind uses to make sense of anything at all.

The rigorous pursuit of mathematics is a formidable gymnasium for the intellect, honing faculties like abstract thought, logical deduction, pattern recognition, and the ability to construct and navigate complex conceptual systems with precision. In grappling with mathematical challenges, one is forced to clarify assumptions, follow chains of reasoning rigorously, and identify the essential structure within intricate problems. This process is not merely about acquiring skills; it is an active engagement with, and a refinement of, cognitive processes. The "aha!" moment of mathematical insight, the aesthetic pleasure derived from an elegant proof, or the deep satisfaction of grasping a complex abstract concept — these subjective experiences offer windows into how minds learn, create meaning, and achieve understanding. The very criteria by which one judges a mathematical argument as "beautiful" or "powerful" may well reflect the mind's innate preferences for order, parsimony, and logical integrity, thus revealing aspects of cognitive aesthetics and the nature of rational intuition.

Beyond its analytical rigor, mathematics is a profoundly creative endeavor, a testament to the boundless imaginative capacity of the human mind. Mathematicians invent new concepts, define novel structures, and explore entire abstract worlds governed by their own internally consistent rules — worlds that often exist initially only within the shared consciousness of the mathematical community. Studying this vast and ever-expanding landscape of mathematical creation allows one to map the potentials and contours of human abstract thought. By understanding the kinds of intricate, often counter-intuitive, yet perfectly logical universes the mind can conceive and find meaningful (from non-Euclidean geometries to transfinite numbers or complex algebraic structures), one learns about the remarkable scope of their own "inner worlds" and their ability to transcend immediate sensory experience. The fact that these abstract constructions, born from the internal logic and creativity of the mind, so often find "unreasonable effectiveness" in describing the external physical world then becomes a fascinating secondary inquiry into the profound and perhaps not yet fully understood resonance between human cognitive architecture and the universe itself. Thus, to study mathematics deeply is to explore the limits and potential of human thought, understanding not just what is, but what can be conceived by the mind, thereby gaining a deeper understanding of one's own remarkable cognitive identity.

Ultimately, whether one views mathematics primarily as a flawless mirror of external reality or a profound reflection of cognitive architecture, the journey into its depths, and into the great scientific narratives it underpins, offers extraordinary rewards. The intricate dance of logic, the pursuit of coherence against daunting complexity, and the breathtaking moments of insight that characterize fields like celestial mechanics or electromagnetic theory are not reserved for a select few. To engage with these topics, at any level of depth, is to connect with one of the most remarkable threads of human intellectual history. It is to witness the astounding capacity of the mind to discern patterns, construct abstract worlds, and make sense of both the cosmos and its own internal landscape. To feel that these ideas are "outside one's reach" is to underestimate the inherent human drive to understand and the diverse paths to appreciation. Exploring these realms is more than an academic exercise; it is an invitation to discover the profound capabilities within one's self and to partake in a shared human treasure. The pursuit enriches understanding not only of the world, but of what it means to be a thinking, comprehending being.

Indices

Terms

- acceleration: 93
- action: 90
- action-at-a-distance: 83, 135
- adequality: 61
- Aha: 18
- algebra: 35, 37
- algorithms: 37
- alternating current: 156
- Ampère's Law: 13, 110
- analysis: 85, 91
- analytic number theory: 88
- angle trisection: 26
- animal electricity: 11
- Annus Mirabilis: 76
- Archimedean spiral: 60
- arithmoi: 24
- atomos: 22
- axioms: 15
- Baku: 18
- Basel problem: 87
- bell curve: 87
- bifilar magnetometer: 106
- binary quadratic forms: 104
- brachistochrone problem: 83
- brachistos: 83
- calculus: 57, 64, 75
- calculus differentialis: 75
- calculus integralis: 75
- calculus of variations: 85, 89
- caloric: 133
- capacitance: 113, 134
- capacitor: 10
- catenary: 83
- catenoid: 89
- characteristic equation: 134
- charge: 109
- Chinese remainder theorem: 36
- chronos: 83
- circuit laws: 113
- coefficients: 109
- combinatorics: 71
- complex function theory: 100
- complex numbers: 89
- conduction current: 129
- congruences: 104
- conic sections: 27
- conjugate: 33
- contact electricity: 12
- convergent: 93
- Coulomb's Law: 110
- cubic: 17
- cuneiform: 16
- curl: 119, 120
- current law: 113
- curvature: 90, 93
- cycloid: 66
- cycloidal cheeks: 67
- cyclotomy: 104
- Daniell cell: 145
- declination: 12, 107
- deferent: 34
- Demiurge: 23
- derivative: 61, 64, 76
- descriptive geometry: 100
- diamagnetism: 56
- diapason: 24
- diapente: 24
- diatessaron: 24
- dielectric: 54, 56
- dielectric constants: 134
- differential equation: 84
- differential triangle: 74
- differentiation: 74
- dimensionality problem: 65
- Diophantine equations: 35
- dip: 12
- dipole: 109
- direct current: 155
- displacement current: 130, 136
- distortionless condition: 161
- divergence: 137
- divergence theorem: 112
- dividing the circle: 104
- dodecahedron: 25
- double-napped cone: 32
- dummy circuit: 154
- duplex: 153
- dyad: 34
- dynamic equilibrium: 92
- effluvium: 8
- elastica: 90
- electricity: 3
- electricus: 8
- electrolysis: 48
- electromotive force: 129, 136
- electron: 3
- electrostatics: 8
- electrotonic state: 124
- electrotyping: 112
- elliptic integrals: 90
- energy: 124
- epicycle: 34
- equant: 34
- equations: 15
- ether: 123
- eudiometer: 52
- Euler angles: 91
- Euler buckling formula: 90
- Euler equations: 90
- Euler's equations: 89
- Euler's equations of rigid body motion: 91
- Euler's formula: 89

- Euler's number: 88
- extrema: 64
- facere: 64
- Faraday effect: 55
- Faraday's Law: 110
- feng shui: 4
- Fermat's Last Theorem: 63
- Fibonacci sequence: 39
- field: 52, 97, 105, 135
- First Law of Planetary Motion: 45
- fluent: 87
- fluents: 74
- fluorescence: 119
- flux: 52, 102, 120
- fluxion: 87
- fluxions: 74
- fluxus: 52
- force: 124
- Fourier analysis: 126
- frequencies: 109
- frequency: 24
- function: 88
- functions: 85
- Fundamental Theorem of Calculus: 74, 75, 76
- galvanic battery: 48
- galvanometer: 51
- galvanoplastics: 112
- gas: 126
- generalized binomial theorem: 76, 78, 87
- generalized coordinates: 95
- geodesics: 89
- gradient: 97, 98
- Green's functions: 118
- group theory: 43, 95
- Hamiltonian: 116, 121
- harmonia: 24
- harmoniai: 24
- harmonic: 98
- Harmonic Law: 45
- harmonics: 93, 98
- heliotrope: 104
- Helmholtz decomposition: 121
- horizon line: 40
- Huygens's Principle: 68
- hydrodynamics: 86
- hyperbola: 28
- Iceland spar: 54
- icosahedron: 25
- idle wheels: 129
- imaginary: 43
- impulse response: 118
- inclination: 107
- incommensurability: 20
- incommensurable: 26
- indivisibles: 65
- inductance: 113
- induction: 41
- inertia: 59
- infinite descent: 20
- infinitesimal: 22, 62
- infinitesimals: 70
- Infinitesimals: 72
- insulators: 53
- integral: 64, 76
- integration: 71, 74
- intensity: 107
- ionosphere: 162
- isochronous: 67
- isoclinic lines: 12
- isoepiphanic: 89
- isogonic lines: 12
- isoperimetric: 89
- kelvins: 119
- kinematics: 124
- kinetic energy: 92
- Lagrangian: 94, 116
- Lagrangian points: 96
- Laplace's equation: 97
- Law of Equal Areas: 45
- law of free fall: 85
- law of quadratic reciprocity: 104
- law of refraction: 83
- least squares: 109
- Legendre polynomials: 98
- lehrfreiheit: 103
- Lenz's law: 51
- lernfreiheit: 103
- Leyden jar: 10
- limit: 94
- limits: 70, 82
- lines of force: 53, 124
- liquid: 126
- loading coils: 155
- loci: 61
- lodestone: 3
- logarithms: 42
- logoi: 24
- logos: 20, 24, 29
- luminiferous ether: 91, 123
- lunar theory: 92
- lunes: 21
- Magnesian stone: 3
- magnet: 3
- magnetic declination: 4
- magnetic flux: 136
- magnetic inclination: 12
- magnetic moment: 105, 106
- magnetism: 56
- magnítis líthos: 3
- marine chronometers: 68
- maximum: 63
- Maxwell distribution: 133
- mean proportionals: 23
- mean speed theorem: 39, 40
- medium: 84
- mesolabium: 32
- method of descent: 96
- method of exhaustion: 26
- method of indivisibles: 65
- method of undetermined multipliers: 95
- Metonic cycle: 22
- minimum: 63
- mirror galvanometer: 151
- model: 15

- modello: 15
- modes: 98
- modus: 15
- moment of inertia: 106
- moments of inertia: 91
- momentum: 92
- monad: 34
- monopole: 109
- multiplier: 95
- mutual inductance: 114
- nappes: 32
- Navier–Stokes equations: 100
- needle telegraph: 111
- Nicol prism: 55
- non-conductors: 53
- normal distribution: 87
- normals: 100
- number theory: 34
- nutation: 91
- octahedron: 25
- Ohm: 131
- one-dimensional wave equation: 93
- optimal control problems: 90
- orthogonal: 136
- parabola: 28, 59
- paramagnetism: 56
- partial differential equation: 93
- Pascal's Law: 69
- Pascal's theorem: 69
- Pascal's Triangle: 70
- Pefsu: 18
- permeability: 13
- permittivity: 134
- perspective transformations: 69
- perturbation functions: 97
- perturbation series: 90
- perturbation theory: 90
- phyllotaxis: 39
- pith ball: 53
- Platonic solids: 25
- polarization: 54
- polynomial equations: 43
- polynomials: 77
- potential function: 118
- potential theory: 98
- precession: 91
- principle of least time: 63, 83
- principle of relativity: 158
- problem of points: 69
- projective geometry: 69
- proto-calculus: 70
- quadratic: 17
- quadratrix: 27
- quadratura: 64
- quadrature: 64
- quartic: 43
- quintic: 43
- ratio test: 94
- rectification: 64, 68
- rectus: 64
- recursive: 88
- regular: 127
- regular convex polyhedra: 25
- regular polygons: 25
- resinous electricity: 9
- resistance: 113
- retentivity: 13
- roots: 134
- scalar: 97
- scalar potential: 109
- scientific method: 28
- Second Law of Planetary Motion: 45
- secular acceleration of the Moon: 96
- semaphore: 142
- series methods: 100
- sexagesimal: 16
- shear stress: 126
- Ship's Part: 18
- sieve of Eratosthenes: 31
- signal retardation: 150
- slide rule: 42
- Sluse's Rule: 70
- solenoids: 13
- solid: 126
- spectroscopy: 113
- spherical harmonics: 98
- squaring the circle: 26, 73
- stadion: 32
- static equilibrium: 92
- stellar parallax: 29
- Stirling's approximation: 87
- Stokes's Theorem: 76, 119, 120
- strain: 126
- strategos: 23
- stress: 126
- subjective relativism: 24
- syllogistic logic: 28
- symbolic algebra: 58
- tangent: 59
- Taylor series: 87
- telegraph equations: 113
- telegrapher's equations: 154
- terrellas: 8
- tetrahedron: 25
- Theorema Egregium: 104
- Third Law of Planetary Motion: 45
- torque: 105
- Torricelli's Trumpet: 71, 72
- transcendental: 78
- transformer: 156
- trichromatic theory: 131
- uniform difform motion: 39
- unit impulse: 118
- vajra: 5
- vanishing point: 40
- vector calculus: 154
- vectors: 97
- vidyut: 5
- vigesimal: 36
- virtual work: 92

- vis viva: 92
- vitreous electricity: 9
- voltage law: 113
- voltaic pile: 12, 48
- vortex: 128
- vortex theory of the atom: 130
- Wallis Product: 71
- wave equation: 89, 94, 113
- weltpolitik: 158
- Wheatstone bridge: 139, 153
- wissenschaft: 103
- work: 92, 124
- zhèn: 6

People

- Abraham de Moivre: 87
- Abu Rayhan al-Biruni: 37
- Adolph Kupffer: 111
- Adrien-Marie Legendre: 98
- Albert Einstein: 158
- Albert Michelson: 158
- Alessandro Volta: 11, 102
- Alexander Graham Bell: 155
- Alexander Neckam: 3
- Alexander von Humboldt: 104
- Alexis Clairaut: 92
- Alfred Vail: 142
- André-Marie Ampère: 13, 50, 123
- Antiphon: 22, 23
- Antoine Arnauld: 73
- Antoine-François Fourcroy: 48
- Antoine-Laurent Lavoisier: 47
- Apollonius: 59
- Apollonius of Perga: 32
- Archimedes: 60
- Archimedes of Syracuse: 30
- Archytas: 24
- Archytas of Tarentum: 23
- Aristarchus of Samos: 30
- Aristotle: 3, 28
- Arthur Cayley: 135
- Arthur Kennelly: 157
- Arthur Zimmermann: 160
- Aryabhata: 37
- Augustin-Jean Fresnel: 63, 123
- Augustin-Louis Cauchy: 94, 126
- Baron Kelvin of Largs: 119
- Benjamin Franklin: 10, 102
- Bernhard Riemann: 94
- Blaise Pascal: 45, 69
- Boethius: 38
- Bonaventura Cavalieri: 65
- Brahmagupta: 37
- Brigham Young: 148
- Brook Taylor: 87
- Callippus: 22
- Carl August von Steinheil: 142
- Carl Friedrich Gauss: 51, 102, 105, 117, 139, 150
- Carl Gustav Jacob Jacobi: 113
- Charles Albert Coffin: 156
- Charles Babbage: 114, 119
- Charles François du Fay: 9
- Charles Tilston Bright: 149
- Charles Wheatstone: 138
- Charles-Augustin Coulomb: 102
- Charles-Augustin de Coulomb: 48
- Charles-François Sturm: 117
- Christiaan Huygens: 45, 67
- Christopher Wren: 81
- Claude-Louis Berthollet: 48
- Claudius Ptolemy: 34
- Colonel Edward Sabine: 117
- Cyrus Field: 149
- Daniel Bernoulli: 86
- Darius Ogden Mills: 148
- David Lloyd George: 159
- Democritus: 22
- Dinostratus: 27
- Diogenes Laertius: 20
- Diophantus of Alexandria: 35
- Edmond Halley: 12, 80
- Edward Bromhead: 115
- Edward Creighton: 147
- Edward Morley: 158
- Edward Orange Wildman Whitehouse: 151
- Edward Routh: 120
- Edward Victor Appleton: 162
- Emil Lenz: 51
- Eratosthenes: 31
- Étienne-Louis Malus: 54
- Euclid: 25, 32, 59
- Euctemon: 22
- Eudoxus: 23, 25
- Eudoxus of Cnidus: 25
- Eutocius of Ascalon: 35
- Evangelista Torricelli: 45, 71
- Ewald Georg von Kleist: 10
- Ezra Cornell: 144
- Félix Savart: 50
- Fibonacci: 39
- Filippo Brunelleschi: 40
- Francesco Maurolico: 41
- Francis Ronalds: 140
- François Arago: 55, 123
- François Viète: 43, 58
- Franz Ernst Neumann: 114
- Franz Neumann: 123
- Franz Ulrich Theodor Aepinus: 11
- Galileo Galilei: 44, 58
- Gaspard Monge: 100
- Georg Simon Ohm: 53
- George Campbell: 161
- George Francis FitzGerald: 154
- George Gabriel Stokes: 119, 134, 149
- George Green: 115
- George Peacock: 114

- George Westinghouse: 156
- Gilles Personne de Roberval: 66
- Girolamo Cardano: 42
- Gottfried Wilhelm Leibniz: 64, 75, 122
- Guglielmo Marconi: 157
- Guillaume de L'Hôpital: 76
- Gustav Kirchhoff: 113
- Hans Christian Ørsted: 13, 50
- Heinrich Hertz: 137
- Hendrik Lorentz: 158
- Henri Poincaré: 158
- Henry Cavendish: 11, 134
- Henry O'Reilly: 146
- Henry Oldenburg: 79
- Henry Savile: 72
- Heraclitus of Ephesus: 20
- Hermann Ludwig Ferdinand von Helmholtz: 121
- Hermann von Helmholtz: 124, 134
- Hero of Alexandria: 33
- Herodotus: 19
- Hesiod: 6
- Hipparchus of Nicaea: 33
- Hippias of Elis: 27
- Hippocrates of Chios: 21
- Hiram Sibley: 146
- Homer: 6
- Humphry Davy: 47, 48
- Iamblichus: 20
- Ibn al-Haytham: 60
- Ignace Gaston Pardies: 73
- Isaac Barrow: 64, 74
- Isaac Newton: 64, 74, 122
- Isambard Kingdom Brunel: 140
- Jakob Bernoulli: 75
- James Clerk Maxwell: 120, 124, 131, 153
- James Faraday: 47
- James Gamble: 147
- James Joule: 119
- James Pierpont Morgan: 156
- James Stirling: 87
- Jane Marcet: 47
- Jean le Rond d'Alembert: 92
- Jean-Baptiste Biot: 50
- Jean-Baptiste Joseph Fourier: 94
- Jean-Victor Poncelet: 100
- Jeptha Wade: 146
- Johann Bernoulli: 75
- Johann Christian Poggendorff: 150
- Johann Peter Gustav Lejeune Dirichlet: 94
- Johannes Kepler: 45
- Johannes Müller von Königsberg: 40
- John Collins: 77
- John Conduitt: 84
- John Couch Adams: 126
- John Dumbleton: 39
- John Frederic Daniell: 144
- John Herschel: 114, 117
- John James Speed, Jr.: 145
- John Lewis Ricardo: 141
- John Michell: 12
- John Napier: 42
- John Pender: 152
- John Toplis: 115
- John Wallis: 71
- John Watkins Brett: 149
- John William Strutt: 135
- Jordanus de Nemore: 39
- Jordanus Nemorarius: 39
- Joseph Barker Stearns: 153
- Joseph Henry: 51
- Joseph Liouville: 117
- Joseph Priestley: 52
- Joseph Sauveur: 93
- Joseph-Louis Lagrange: 94
- Jost Bürgi: 42
- Kaiser Wilhelm II: 158
- Katherine Mary Dewar: 131
- King Hiero II of Syracuse: 30
- Leon Battista Alberti: 40
- Leonardo of Pisa: 39
- Leonhard Euler: 85
- Leucippus: 22
- Liu Hui: 36
- Lodovico Ferrari: 43
- Lord Kelvin: 119
- Lord Rayleigh: 135
- Louis Navier: 100
- Louis-Bernard Guyton de Morveau: 48
- Luigi Galvani: 11, 102
- Marin Mersenne: 45
- Marshall Lefferts: 148
- Matthew Fontaine Maury: 149
- Menaechmus: 27, 32
- Meton of Athens: 22
- Michael Faraday: 14, 46, 47, 49, 120, 136
- Michel Chasles: 117
- Mihajlo Pupin: 161
- Mikhail Ostrogradsky: 112
- Moritz von Jacobi: 112
- Muhammad ibn Musa al-Khwarizmi: 37
- Muḥammad ibn Mūsā al-Khwarizmi: 58
- Niccolò Tartaglia: 42, 59
- Nicolas Fatio de Duillier: 84
- Nicolaus Copernicus: 41
- Nicole Oresme: 40, 58
- Nicomachus of Gerasa: 33
- Nikola Tesla: 156

- Oliver Heaviside: 138, 152
- Oliver Lodge: 154, 157
- Omar Khayyam: 38
- Ostilio Ricci: 44
- Otto von Guericke: 9
- Pappus of Alexandria: 35
- Parmenides of Elea: 21
- Paul Guldin: 65
- Pavel Schilling: 111, 139
- Peter Mark Roget: 139
- Petrus Peregrinus de Maricourt: 7
- Philo Farnsworth: 161
- Philolaus: 21
- Pierre de Fermat: 45, 60
- Pierre-Simon Laplace: 96, 102
- Pieter van Musschenbroek: 10
- Plato: 3, 20, 23
- Pliny the Elder: 3
- Plutarch: 30
- Porphyry: 20
- Ptolemy: 30
- Ptolemy III Euergetes: 32
- Ptolemy IV Philopator: 32
- Pythagoras: 19, 20
- Qin Jiushao: 36
- Rafael Bombelli: 44
- Regiomontanus: 40
- René Descartes: 45, 58
- René-François de Sluse: 70
- Richard Swineshead: 39
- Robert Boyle: 9, 47
- Robert Hooke: 79, 81
- Robert Murphy: 116
- Robert Recorde: 42
- Rudolf Kohlrausch: 110
- Saint Francis of Paola: 45
- Samuel Finley Breese Morse: 141
- Samuel Thomas von Sömmerring: 111
- Scipione del Ferro: 42
- Siméon-Denis Poisson: 101, 102
- Simon Stevin: 42
- Simplicius: 21
- Socrates: 24
- Stephen Gray: 9
- Sylvestre François Lacroix: 115
- Thales: 19
- Thales of Miletus: 3
- Theaetetus: 24
- Theano: 20
- Theudius of Magnesia: 28
- Thomas Bradwardine: 39
- Thomas Edison: 155
- Thomas Heaviside: 138
- Thomas Hobbes: 45, 72
- Thomas Young: 131
- Tycho Brahe: 45
- Vitruvius: 38
- Werner von Siemens: 155
- Wilhelm von Humboldt: 105
- Wilhelm Weber: 105, 123, 139, 150
- Willebrord Snellius: 63
- William Cavendish: 134
- William Fothergill Cooke: 139
- William Gilbert: 8
- William Herschel: 133
- William Heytesbury: 39
- William Hopkins: 117, 120
- William Nicol: 55
- William Oughtred: 42
- William Preece: 154
- William Rowan Hamilton: 116, 121
- William Thomson: 117, 121, 124, 130, 134
- Zeno of Citium: 29
- Zeno of Elea: 21

Other

- Abbasid Caliphate: 7
- Academy of Sciences: 56
- Adams Prize: 126
- Alert: 159
- Allgemeine Elektricitäts-Gesellschaft: 159
- American Civil War: 148
- American Marconi: 159
- American Telegraph Company: 146, 148
- American Telephone and Telegraph Company: 160
- Anglo-American Telegraph Company: 152
- Anglo-Danish Telegraph Company: 152
- Apache raids: 147
- Archaic: 19
- Ark of the Covenant: 6
- Atlantic Telegraph Company: 148
- Atlantis: 23
- Austro-German Telegraph Union: 150
- Baghdad Battery: 5
- Baltimore & Ohio Railroad: 143
- Bell Telephone Company: 155
- Bell Telephone Laboratories: 161
- British Association for the Advancement of Science: 131, 149
- California statehood debate: 147
- Cataract Construction Company: 156
- Cavendish Professor: 135
- Chancellor of the Exchequer: 159
- Chemical Revolution: 47
- Chief Night Operator: 153
- Civil War: 72
- Commentarii: 87
- Count of the Empire: 99
- Crelle's Journal: 118
- Crusades: 7
- Curator of Experiments: 79
- Delian problem: 23, 26, 27, 32
- Eastern & Associated Telegraph Companies: 157
- École Polytechnique: 99
- Edison General Electric: 156
- Electric Telegraph Company: 141
- English Mechanic: 153
- Erie & Michigan: 146
- Ferme Générale: 48
- First Consul: 99
- French Revolution: 48
- General Electric: 156
- Göbekli Tepe: 16
- Göttingen Seven: 108
- Great Inequality: 97
- Great Northern Telegraph Company: 152
- Great Plague of London: 76
- Great Pyramid: 5
- Great Recoinage: 82
- Great Western Railway: 139
- Greek Dark Age: 19
- HMS Agamemnon: 149
- House of Wisdom: 7, 37
- Humboldtian model: 101
- Illinois & Mississippi: 147
- Imperial Wireless Chain: 159
- Industrial Revolution: 130
- Institut de France: 99
- International Niagara Commission: 156
- International Radiotelegraph Conference: 159
- Islamic Golden Age: 37
- Journal of the Royal Institution: 49
- Khujut Rabu: 4
- Lake Shore: 146, 147
- Late Bronze Age: 18
- Lefkandi: 19
- Lucasian Professor of Mathematics: 77
- Magnetic Crusade: 117
- Magnetic Telegraph Company: 145
- Magnetic Union: 107, 117, 139
- Magnetischer Verein: 107
- Marconi Scandal: 159
- Master of the Mint: 82
- Mathematical Tripos: 115
- Maxwellians: 154
- Maya: 36
- Member of Parliament: 82
- Merton College: 39
- Military Academy: 96
- Minister of the Interior: 99
- Minister of the Marine: 100
- Mjölnir: 6
- Nauen Transmitter Station: 159
- Nazca Lines: 16
- Neoplatonic: 34
- New Kingdom: 19
- New World: 148
- New York & Mississippi Valley: 147

- New York & Mississippi Valley Printing Telegraph Company: 146
- Newcastle Literary and Philosophical Society: 153
- Normal School: 100
- Nottingham Subscription Library: 115
- O'Reilly v. Morse: 146
- Ohio Indiana & Illinois: 146
- Ohio, Indiana & Illinois: 147
- Old World: 148
- Olmecs: 4
- Order of Minims: 45
- Oxford Calculators: 39
- Pacific Telegraph Act: 148
- Panic of 1837: 142
- Pascaline: 69
- Philosophical Magazine: 153
- Pneumatic Institution: 48
- Polytechnic School: 100
- Pony Express: 147
- Prize Fellow: 121
- Professorship of Natural Philosophy: 128
- Radio Act: 159
- Radio Corporation of America: 161
- Reconquista: 7
- Reign of Terror: 99
- Renaissance: 38
- Republic of Letters: 62, 73
- Revolutions of 1848: 110
- Rochester: 147
- Rochester Telegraph Company: 146
- Room 40: 160
- Royal Engineers: 149
- Royal Institution: 47
- Saint Paul's Cathedral: 81
- Savilian Professor of Geometry: 72
- Second Wrangler: 117
- Senior Wrangler: 119
- Siemens & Halske: 155
- Smith's Prize: 120
- Society of Telegraph Engineers: 153
- Song: 36
- Song Dynasty: 4
- Sophists: 22
- Southern Telegraph Company: 146
- Speedwell Iron Works: 142
- SS Great Eastern: 152
- Stoic: 29
- Stonehenge: 16
- Strait of Dover: 149
- Telcon: 152
- Telefunken: 159
- Telegraph Act: 152
- Telegraph Bill: 143
- Telegraphic Plateau: 149
- Temple of Hathor: 5
- The English Mechanic: 154
- The Great War: 138, 159
- The Telegraphic Journal: 154
- Thomson-Houston Electric Company: 156
- Titanic: 159
- Trinity: 80
- Tripos: 117, 120, 135
- United Fruit Company: 161
- University of Berlin: 101
- Uraniborg: 45
- USS Niagara: 149
- Utah war: 147
- War of the Currents: 155
- Warden of the Royal Mint: 82
- Wardenclyffe: 160
- Western Union: 147
- Wireless Telegraphy Act: 159
- World's Fair: 156
- Wranglers: 120
- Yuan: 36
- Zimmermann Telegram: 160

Publications

- A Chart of the North Atlantic Ocean: 149
- A Dynamical Theory of the Electromagnetic Field: 132
- A New Method for Maxima and Minima…: 75
- A treatise of celestial mechanics: 99
- Achilles and the Tortoise: 21
- Algebra: 44
- Almagest: 34
- Amos Dettonville: 70
- An Attempt at a Theory of Electricity and Magnetism: 11
- An Essay on the Application of Mathematical Analysis to the Theories of Electricity and Magnetism: 115
- Analysis of the Infinitely Small for the Understanding of Curves: 76
- Analytical Mechanics: 94
- Analytical Society: 114
- Arithmetic of Infinites: 71
- Arithmetica: 35
- Arithmetical Investigations: 104
- Arrow: 21
- Book of Changes: 6
- Book of Exodus: 6
- Book of Job: 6
- Book of Mountains and Seas: 6
- Collection: 35
- Comparing Electromotive Forces: 153
- Conics: 32
- Conversations on Chemistry: 47
- Critias: 23
- De naturis rerum: 3
- Dichotomy: 21
- Direct and Indirect Methods of Incrementation: 87
- Dream Pool Essays: 4
- Duplex Telegraphy: 154
- Eddas: 6
- Elements: 28, 34, 59, 72
- Elements of Chemistry: 47
- Elements of Geometry: 21
- Essay on Conics: 69
- Experimental Researches in Electricity: 52
- Experiments on the effect of a current of electricity on the magnetic needle: 13
- Foundations of Differential Calculus: 85
- Foundations of Integral Calculus: 85
- General Investigations of Curved Surfaces: 104
- Geometrical Lectures: 74
- Geometry: 45
- Gospel of John: 29
- History of Fishes: 81
- Introduction to Arithmetic: 34
- Introduction to Plane and Solid Loci: 61
- Introduction to the Analysis of the Infinite: 85
- Ion: 3
- Journal of Egyptian Archaeology: 18
- Letters on the Magnet: 7
- Leviathan: 72
- Limit: 94
- Lore and Science in Ancient Pythagoreanism: 20
- Master Lü's Spring and Autumn Annals: 4
- Mathematical Treatise in Nine Sections: 36
- Measurement of a Circle: 31
- Mechanics, or the Science of Motion Presented Analytically: 88
- Metaphysics: 28
- Method of Chemical Nomenclature: 48
- Method of Fluxions: 75, 76, 77
- Method of Fluxions and Infinite Series: 74
- Metrica: 33
- Micrographia: 79
- Moscow Mathematical Papyrus: 17
- Naturalis Historia: 3
- New Theory about Light and Colors: 78
- On a Hidden Geometry and the Analysis of Indivisibles and Infinites: 75
- On Analysis by Equations with Infinitely Many Terms: 77
- On Architecture: 38
- On Faraday's Lines of Force: 124
- On Governors: 133
- On Nature: 21
- On Painting: 40
- On Physical Lines of Force: 128
- On Piety: 20

- On the Conservation of Force: 121, 124
- On the Equilibrium of Planes: 31
- On the Extra Current: 154
- On the Magnet: 8
- On the Motion of Bodies in Orbit: 81
- On the propagation of electricity in wires: 113
- On the Revolutions of the Heavenly Spheres: 41
- On the Speed of Signalling through Heterogeneous Telegraph Circuits: 154
- On the Sphere and Cylinder: 31
- Opticks: 82
- Organon: 28
- Physics: 28
- Pingzhou Table Talks: 4
- Quadrature of the Parabola: 64
- Reprint of Papers on Electrostatics and Magnetism: 118
- Researches, Chemical and Philosophical; Chiefly Concerning Nitrous Oxide: 48
- Results from the Observations of the Magnetic Union: 107
- Rhind Mathematical Papyrus: 17, 22
- Rigveda: 5
- Rules of the Cord: 37
- Seventh Letter: 23
- Ten Commandments: 6
- The Book of Calculation: 39
- The Compendious Book on Calculation by Completion and Balancing: 37
- The Correctly Established Doctrine of Brahma: 37
- The Doctrine of Chances: 87
- The General Laws of Induced Electric Currents: 114
- The Giza Power Plant: 5
- The Great Art: 43
- The Key to Mathematics: 42
- The Mathematical Principles of Natural Philosophy: 81
- The Method: 65
- The Method of Mechanical Theorems: 31
- The Nine Chapters on the Mathematical Art: 36
- The Pendulum Clock: 67
- The Sceptical Chymist: 47
- The Tenth: 42
- The Whetstone of Witte: 42
- Theaetetus: 24
- Theogony: 6
- Theory of Heat: 133
- Timaeus: 23
- Treatise on Differential and Integral Calculus: 115
- Treatise on Dynamics: 92
- Treatise on Electricity and Magnetism: 135, 153
- Treatise on the Arithmetical Triangle: 70
- Two Books of Arithmetic: 41
- Works and Days: 6

www.ingramcontent.com/pod-product-compliance
Lightning Source LLC
LaVergne TN
LVHW010326070526
838199LV00065B/5669